JIT Implementation Manual

The Complete Guide to
Just-in-Time Manufacturing

Second Edition

Volume 6

JIT Implementation Manual

The Complete Guide to
Just-in-Time Manufacturing

Second Edition

Volume 6
JIT Implementation Forms and Charts

The software mentioned in this book is now available for download on our Web site at: http://www.crcpress.com/e_products/downloads/default.asp

HIROYUKI HIRANO

CRC Press
Taylor & Francis Group
Boca Raton London New York

CRC Press is an imprint of the
Taylor & Francis Group, an **informa** business

A PRODUCTIVITY PRESS BOOK

Originally published as *Jyasuto in taimu seisan kakumei shido manyuaru* copyright © 1989 by JIT Management Laboratory Company, Ltd., Tokyo, Japan.

English translation copyright © 1990, 2009 Productivity Press.

CRC Press
Taylor & Francis Group
6000 Broken Sound Parkway NW, Suite 300
Boca Raton, FL 33487-2742

© 2009 by Taylor & Francis Group, LLC
CRC Press is an imprint of Taylor & Francis Group, an Informa business

No claim to original U.S. Government works
Printed in the United States of America on acid-free paper
10 9 8 7 6 5 4 3 2 1

International Standard Book Number-13: 978-1-4200-9032-1 (Softcover)

Visit the Taylor & Francis Web site at
http://www.taylorandfrancis.com

and the CRC Press Web site at
http://www.crcpress.com

Contents

Volume 3

Volume 4

Chapter 16

JIT Forms

This chapter represents a comprehensive collection of JIT forms and other tools that will come in handy for promoting the JIT factory revolution. These forms are divided into five categories: overall management, waste-related forms, 5S-related forms, engineering-related forms, and JIT introduction-related forms. The forms are provided both as filled-in examples and as blank forms that can be removed from the binder and photocopied for use in the factory.

The table in Figures 16.1 and 16.2 list titles and brief descriptions of the JIT forms contained in the chapter.

		Title	Description
Overall Management	1	Diagnostic List for JIT Management	Use this form to enter diagnostic data and describe current conditions, goals, and other data relevant to management that supports the JIT factory revolution.
	2	5-Level List of JIT Manufacturing Functions, with radar chart supplement	This list organizes JIT production's 13 functions into 5 levels. We can then use a radar chart to illustrate their interrelations.
Waste-related Forms	3	Arrow Diagrams	We use arrow diagrams to analyze the flow of goods in the factory and remove the major forms of waste.
	4	Summary Chart of Flow Analysis	This chart lists the "before improvement" and "after improvement" status of items analyzed in arrow diagrams.
	5	Operations Analysis Table	We use this table for describing and analyzing the entire series of operations in any part of the factory, from raw materials to finished products.
	6	Waste-finding Check-lists (workshop-specific and process-specific)	These two lists, one for workshops and the other for processes, help us discover where waste has been hidden.
	7	5W1H Sheet	This analysis sheet helps us find the true causes of problems that occur in the workshops.
5S-related Forms	8	5S Checklist (for factories)	Used for checking how well the 5S's are enforced throughout the factory. Can also be used to check on 5S conditions at outside supplier companies.
	9	5S Checklist (for workshops and offices), with radar chart supplement	Data from these separate 5S checklists for manufacturing workshops and clerical or administrative offices can be used to draw radar charts showing relative strengths and weaknesses, which are useful as score-keeping tools for 5S contests.
	10	5S Memos	These include 5S maps and other means of indicating 5S conditions.
	11	Red Tags	These are the red tags used in the "red tag strategy."
	12	Red Tag Strategy Report Form	The results of red tag campaigns are entered on this form.
	13	Unneeded Inventory Items List and Un-needed Equipment List	Working from red tag campaign results, this list is for entering the types of unneeded inventory and the disposal method for each type.
	14	Cleanliness Inspection Checklist	This provides a tabular form for entering check point descriptions when inspecting cleanliness conditions.
	15	5-Point Cleaned-Up Checklist	This checklist has five levels of "cleaned-up" status for each of three "S" categories.
	16	Display Boards	These signboards show where certain items are to be placed temporarily.

Figure 16.1   List of JIT Forms.

		Title	Description
Engineering-related Forms	17	PQ Analysis List and Charts	Used to estimate output quantities of products and components.
	18	Process Route Diagram	Clarifies the relationship between the flow of goods and the use of equipment.
	19	Line Balance Analysis Charts	Table for analyzing balance in assembly line operations.
	20	Cooperative Operation Confirmation Charts	Table for confirming cooperative operations on assembly lines, etc.
	21	Vendor Delivery Evaluation Charts	Charts for comprehensive analysis of delivery methods used for purchases and subcontracted goods.
	22	JIT Delivery Efficiency Lists	Lists for evaluating delivery efficiency of purchased and subcontracted goods.
	23	Flexible Production Training Schedule	Easy-to-read report of conditions and progress in multiple skills training.
	24	Flexible Production Map	Easy-to-read description of conditions and progress in multiple skills training.
	25	Production Management Boards	Tool for comparing estimated production schedule with actual production on an hour-to-hour basis.
	26	Model and Operating Rate Trend Charts	Chart for confirming changeover-related needs.
	27	Public Changeover Timetables	Presents an overview of changeover operations.
	28	Changeover Improvement Lists	For in-depth study of improvement items and confirmation of progress in making improvements.
	29	Changeover Work Procedure Analysis Charts	Elucidates minor operations and points toward improvements.
	30	Changeover Results Tables	Promotes better understanding of current changeover operations.
	31	5S Checklist for Changeover	For use in checking on 5S maintenance as it relates to changeover.
	32	*Poka-Yoke*/Zero Defects Checklist	For flushing out causes of defects and setting improvements into the proper sequence.
	33	Parts-Production Capacity Work Table	Shows the basic times and other performance-related data for processing of each part.
	34	Standard Operation Combination Charts	Helps us find the most efficient combination of human work and machine work.
	35	Summary Table of Standard Operations	Summarizes key points and critical factors in standard operations.

Figure 16.2   List of JIT Forms.

		Title	Description
JIT Introduction-related Forms	36	Work Methods Table	Used for giving advice or training to equipment operators.
	37	Standard Operation Forms	Shows correct operations, within the cycle time.
	38	JIT's Ten Commandments	Presents the ten most basic concepts and precepts of JIT production.
	39	Improvement Memos	Provides a handy form for memos regarding JIT improvement activities.
	40	List of JIT Improvement Items	Lists improvement items and keeps track of progress.
	41	Improvement Campaign Planning Sheet	For checking on the progress of large or long-term improvement themes.
	42	Improvement Result Charts	Enables easy, visual "before" and "after" comparison of improvements.
	43	Weekly Report on JIT Improvements	Form for weekly reports from affiliated companies or factories.
	44	JIT Leader's Report	Form for JIT leaders to use in giving advice.

Figure 16.2 (continued)

Overall Management

JIT Management Diagnostic List

Application

As an aide to promoting the JIT factory revolution, this list helps describe the overall company organization and provides a form for setting and listing JIT improvement goals. As such, this list (see Figures 16.3 to 16.5) can also be useful for managing outside orders.

Main sections of form

1. *Total value added.* This is the remainder obtained by subtracting total expenditures from the total value of output.
2. *Inventory assets.* These assets are divided into three categories: products, in-process inventory, and materials and purchased parts.
   a. *Products:* The value of product inventory indicates the company's overall strength in sales, manufacturing, and distribution. The lower the product inventory value, the better.
   b. *In-process inventory:* This indicates how strong the company is in terms of maintaining a streamlined flow of goods. The smaller the in-process inventory value, the better.
   c. *Materials and purchased parts:* This value figure shows how efficient the company is in purchasing. Again, the smaller the value, the better.
3. *Production techniques.* This section is for entering which kind of production method is being used (lot production or flow production), along with a short description and comments.

JIT Management Diagnostic List

Date:

Company:	Capitalization:
Address:	Telephone:
Major products:	

Management and finance

	Item	Description		Current condition	Goals, comments, etc.
1	Value of sales (annual)		($)		
2	Operating profit	(1)	($)		
		(2) Ratio	(%)		
		(3) Per employee	($)		
3	No. of employees	(1) Indirect No. of men No. of women	($)		

Figure 16.3   Example of a JIT Management Diagnostic List.

JIT Management Diagnostic List

Date:

Company:	**Capitalization:**
Address:	**Telephone:**
Major products:	

Management and finance

	Item	Description		Current condition	Goals, comments, etc.
1	**Value of sales (annual)**		($)		
2	**Operating profit**	(1)	($)		
		(2) Ratio	(%)		
		(3) Per employee	($)		
3	**No. of employees**	(1) Indirect No. of men No. of women	($)		
		(2) Indirect No. of men No. of women	($)		
4	**Labor union?**	YES ● NO	If YES, does union belong to a parent organization?		
5	**Value of net productivity**	Per share	($)		
6	**Debts**		($)		
7	**Inventory assets**	(1) Product value ($) ÷ ratio (times/month)			
		(2) In-process inventory value ($) ÷ ratio (times/month)			
		(3) Materials value ($) ÷ ratio (times/month)			
		(4) Total value ($) ÷ ratio (times/month)			
8	**Plant investment**	Investment value ÷ Depreciation ratio			
9	**Building space**	(1) Factory (m2)			
		(2) Warehouse (m2)			
		(3) Office (m2)			
		(4) Other (m2)			
10	**Outside contractors and vendors**	(1) No. of outside contractors or vendors			
		(2) Value of outside orders ($) or purchases ($)			

Figure 16.4   JIT Management Diagnostic List (Value-Added and Inventory Assets).

Factory			

	Item	Description	Current condition	Goals, comments, etc.
1	**Shift system**	No. of shifts		
2	**Absentee rate**	(%)		
3	**Overtime hours per month**	Total overtime per person		
4	**Product-specific lead-time**	(1) Product 1: (No. of days) ($)		
		(1) Product 2: ($)		
5	**Process-specific lead-time**	(1) Purchasing lead-time ($) (No. of days)		
		(2) Subcontracting lead-time ($) (No. of days)		
		(3) Processing lead-time ($) (No. of days)		
		(4) Assembly lead-time ($) (No. of days)		
6	**Production techniques**	(1) Job shop or flow shop?		
		(2) Multi-machine operations or multi-process operations?		
		(3) Lot production or one-piece flow production?		
		(4) Sitting while working or standing while working?		
		(5) Downstream inspection or independent inspection?		
		(6) Single skills or multiple skills?		
		(7) Process-specific pitch or product-specific pitch?		
		(8) Lots of waste or not much waste?		
		(9) Lots of inventory or not much inventory?		
		(10) Lots of defects or not many defects?		
7	**5S's (*seiri, seiton, seiso, seiketsu, shitsuke*)**	Have the 5S's been established?		

Overall evaluation	General comments

Figure 16.5 JIT Management Diagnostic List (Production Techniques).

Five Stages of JIT Production and JIT Radar Charts

Application

Use this to evaluate how well the factory is doing in terms of JIT's 13 main functions. The radar chart enables us to gain an immediate grasp of the company's relative strengths and weaknesses in these 13 functions.

Main sections of form:

1. *First Level (Little League).* This level is typical of the struggling, money-losing company whose survival is in doubt.
2. *Second Level (Junior Varsity).* Companies at this level are managing to survive, for the time being at least.
3. *Third Level (Varsity).* Companies at this level are doing just well enough to not be ashamed to host factory tours.
4. *Fourth Level (Minor League Pro).* At this level, companies are doing well enough to take pride in being able to teach other companies a thing or two.
5. *Fifth Level (Major League Pro).* These top-ranking companies truly have what it takes to survive into the 21st century.

The structure of the JIT production system is illustrated in the following diagram (see Figures 16.6 to 16.11).

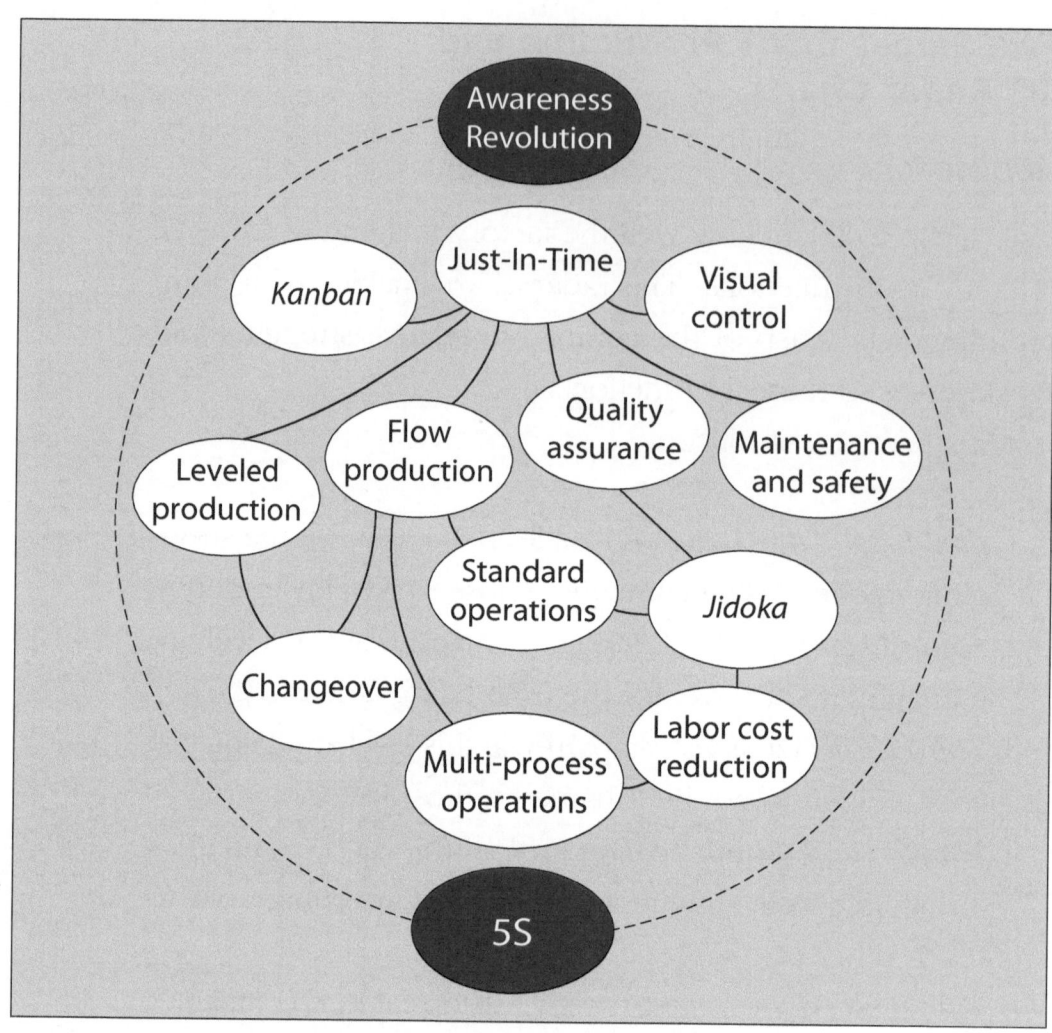

Figure 16.6   Structure of the JIT Production System.

1. Awareness Revolution

 Almost everyone in the factory thinks we are still in the era of large-lot production and that the current way of making things is about as good as it can be.

 To almost everyone in the factory, putting customer service first is a foreign concept. Instead, the emphasis is on facilitating production.

 Some of the people have at least an intellectual understanding of the need to put the customer first. However, this new thinking is not reflected in the factory, which clings to its old ways.

 Almost everyone knows how important it is to put the customer first, and the factory itself is gradually beginning to reflect this.

 The "customer first" concept has penetrated to every corner of the factory. Whenever a problem or abnormality occurs, people get to work at once to make a corrective improvement.

2. The 5S's

 Everything in the factory is lying around in disorganized heaps. In fact, even the people who habitually use certain things often have no idea where the things are or how many of them are in stock.

 Everything appears to be a mess, but somehow the people using the things usually know where to find them.

 White lines demarcate work areas and paths; and tools, in-process inventory, and machines appear to be put into some kind of logical order.

 Tools, in-process inventory, and machines are marked with location indicators, and the floor and machines are kept clean. The causal observer would believe that everything is clean and well-organized.

 Things are marked so that anyone can tell what goes where and in what amount, everything is kept clean, and devices have been developed to help prevent things from getting dirty in the first place.

3. Flow Production

 Equipment is laid out job-shop style, large lots accumulate near various machines and operators, and each process moves at its own pitch.

 Operators are trying to handle smaller lots, but since the equipment layout is still in the job-shop style, production relies heavily on the conveyance system.

 The equipment has been rearranged for in-line layout, but production flow is limited to the single-process small-lot method.

 Production has switched to one-piece flow based on hand-conveyance, single-process operations, and in-line equipment layout.

 Operators are standing while working and carrying out multi-process operations based on one-piece flow production.

Figure 16.7   JIT Production's Five Stages of Development.

4. Multi-Process Operations

 Workers do not want to change. They insist that they are only able (or willing) to do the job they are currently doing.

 Operators are carrying out "caravan" operations, but operators at adjacent processes do help each other out now and then to improve the flow of goods.

 Processes are lined up to facilitate the flow of goods, and adjacent operators regularly help each other out.

 All operators are standing while working. Each operator is able to handle about half of the processes in the cell.

 Processes are lined up to facilitate the flow of goods. Each operator is able to handle all processes in the cell.

5. Labor Cost Reduction

 Operators do not move efficiently and there are clearly more workers in the factory than necessary.

 Overstaffing is not so evident. Everyone in the production line always does the same tasks, and the balance of operations is poor.

 Job duties differ only from product to product and the balance of operations is basically OK.

 Job duties are adaptable to changes in required output.

 Operators are trained flexibly and can work anywhere on the line. The number of workers is kept to the minimum needed to produce the required output.

6. *Kanban*

 Workpieces are pushed downstream and processes are arranged in no apparent order.

 Push production still prevails, but things are generally organized into specified temporary storage areas.

 Things are kept in specified places and specified amounts, and ways are being found to switch from push production to pull production.

 Downstream processes are withdrawing *kanban* from upstream processes.

 Withdrawal of *kanban* from downstream processes is being combined with ongoing improvement activities.

Figure 16.7 (continued)

7. Visual Control

 No one can tell when an abnormality occurs, so the production line keeps on going.

 No one can tell when an abnormality occurs, but they are eventually discovered and corrected.

 The people directly involved can tell when things are normal and when they are abnormal, and they respond (sooner or later) to abnormalities.

 Everyone can tell when things are normal and when they are abnormal, and they respond (sooner or later) to abnormalities.

 Everyone can tell when things are normal and when they are abnormal, and they respond at once to abnormalities.

8. Leveled Production

 Each product model has only one run per month, and each process moves at its own pitch.

 Each product model has only two runs per month, and each process moves at its own pitch.

 Each product model has four runs per month (one per week), and some synchronization of processes has been achieved.

 Monthly production schedules are divided into daily production runs, and in-line production has been established with specific cycle times.

 Fully-leveled production has been established, and the cycle time sets the rhythm for the entire factory.

9. Changeover

 Only one or two changeovers per month regardless of customer needs. Changeover times can be as long as half a day.

 People are conscious of the need to orient changeovers toward serving customer needs.

 Changeover teams have been formed to improve external changeover, etc.

 Changeover times have been shortened significantly.

 No changeover operation takes more than three minutes and all are done within the cycle time.

Figure 16.7 (continued)

10. Quality Assurance

 Lots of defective products get shipped, resulting in lots of customer complaints.

 Defects occur, but a strict final inspection process keeps customer complaints low.

 Improvement teams have been formed and use inspection data in responding to defects.

 Defects are detected before being passed to the next process by operators who perform independent inspection and improvements.

 Jidoka and *poka-yoke* devices have been developed to build in quality at each process and to detect defects at their source to prevent occurrence or recurrence.

11. Standard Operations

 Operation methods are left up to the operators, who depend on their experience and "instincts" to do the job correctly.

 Operators tend to perform their tasks in similar ways, but there is no attempt at improving standardization.

 Process-specific standards have been established and are generally followed.

 Systematic production standards are followed at each process, but there is no attempt to improve them.

 Standard operations are well-defined, followed completely, and constantly improved upon.

12. *Jidoka*

 All operations are done either manually or by expensive large-lot processing equipment.

 Operations are done by machines but always with human assistance.

 Workers have been separated from the machines. Machines producing defective goods must be turned off manually.

 Workers have been separated from the machines that start turning out defective products.

 Separation of workers and *jidoka* have been successfully extended to the assembly line.

Figure 16.7 (continued)

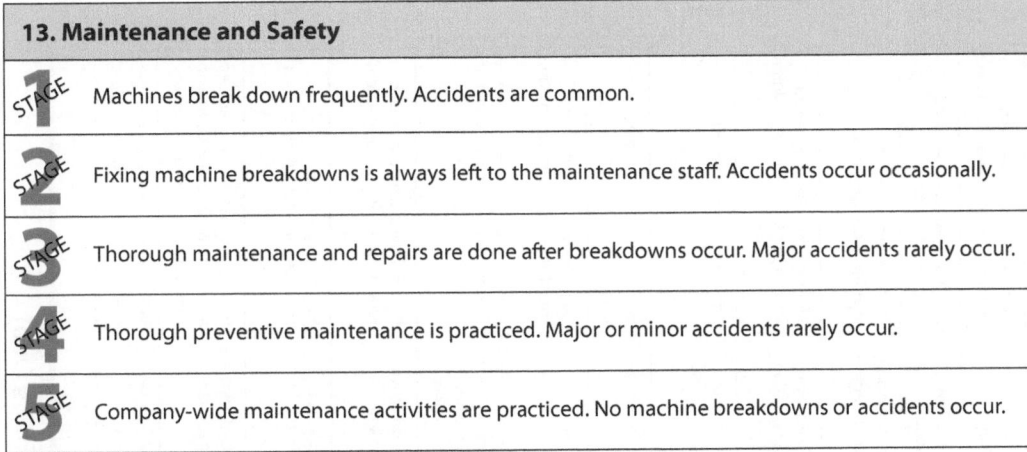

13. Maintenance and Safety
STAGE 1 Machines break down frequently. Accidents are common.
STAGE 2 Fixing machine breakdowns is always left to the maintenance staff. Accidents occur occasionally.
STAGE 3 Thorough maintenance and repairs are done after breakdowns occur. Major accidents rarely occur.
STAGE 4 Thorough preventive maintenance is practiced. Major or minor accidents rarely occur.
STAGE 5 Company-wide maintenance activities are practiced. No machine breakdowns or accidents occur.

Figure 16.7 (continued)

List of JIT's Major Functions and Their Five Stages of Development

Name of workshop:
Ranked by:
Date:

No.	Level (baseball rank) / Function	First Level (Little League)	Second Level (Junior Varsity)	Third Level (Varsity)	Fourth Level (Minor League Pro)	Fifth Level (Major League Pro)
1	Awareness Revolution	Large-scale mass production for maximum output	Product-out orientation	Market-in orientation, but not implemented in each workshop	Service orientation, with service-oriented workshops	Service orientation implemented at each process factory-wide
2	The 5S's	Hard for anyone to tell what goes where and when	Hard for visitors to tell what goes where and when, but workers know	Factory uses outlining and classification for visual control	Good indicators and clean, neatly organized factory	Clean, neatly organized with mess-prevention measures in force
3	Flow Production	Job-shop layout, geared for large-lot production	Job-shop layout, geared for small-lot production	In-line layout, small-lot flow at and between processes	In-line layout, one-piece flow at and between processes	Full multi-process operations with one-piece flow
4	Multi-process Operations	Unquestioned support for single-skill, single-process operations	Caravan-style cooperative operations	Flow-based cooperative operations	About halfway toward achieving smooth multi-process operations	Smooth and complete multi-process operations
5	Labor Cost Reduction	Wasteful motion and too many workers	Fixed job assignments and poor balance	Fixed job assignments, but different for each model, slightly better balance	About halfway toward achieving smooth multi-process operations	Flexible job assignments, with narrow variation in output volume
6	Kanban	Push production, with retained inventory all over the place	Push production, with organized storage sites for in-process inventory	Pull production, with fixed locations and fixed volumes	Flexible job assignments, with wide variation in output volume	Kanban and improvements
7	Visual Control	Abnormalities often occur and only create confusion	Abnormalities often occur and are usually resolved in some way	Supervisors can tell when an abnormality occurs	Pull production, with kanban	Immediate action is taken to resolve abnormalities
8	Level Production	Once-a-month production schedule, processes have own rhythms	Twice-a-month production schedule, each process has its own rhythm	Weekly production schedule, overall line has some kind of common rhythm	Anyone can tell when an abnormality occurs	Completely level production, overall line has a common rhythm
9	Changeover	Monthly changeover, requires half a day each time	People are aware of changeover needs	Changeover teams and improvements made in some workshops	Daily production schedule, overall line has a common rhythm. Single-operation changeovers	Changeovers are within cycle times
10	Quality Assurance	Factory ships defective products and deals with customer complaints	Defective products are sorted out at final inspection and not shipped	Factory produces defective products but passes information to reduce defects	Processes do not send defectives downstream (independent inspection)	Factory builds quality in at each process (at-the-source inspection)
11	Standard Operations	Operation procedures are generally left up to each operator	Operation procedures are vaguely standardized in roughly the same order	Standard operations implemented for individual processes	Standard operations planned but not fully implemented	Standard operations and improvements fully implemented
12	Human Automation	All processes require manual assistance, lots of large-lot equipment	Some automation, but operators are always present while machines work	Human and machine work separated; machines sometimes make defective items	Human and machine work separate, but machines sometimes make defectives	Human and machine work are separate, with no defectives, and with some human automation devices
13	Maintenance and Safety	Lots of breakdowns and numerous accidents per year	Factory uses maintenance specialists but has occasional accidents	Factory has follow-up maintenance and no major accidents	Factory has preventive maintenance and is almost accident-free	Factory has company-wide preventive maintenance and no accidents

Figure 16.8   List of JIT's Major Functions and Their Five Stages of Development (Example of its Use).

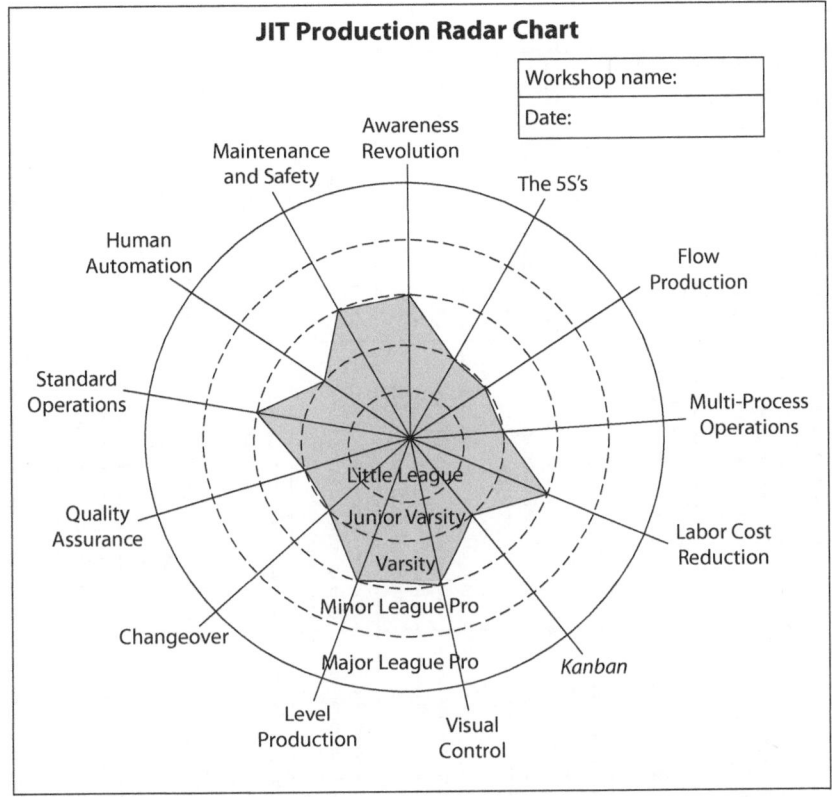

Figure 16.9   Example of a JIT Production Radar Chart.

List of JIT's Major Functions and Their Five Stages of Development

Name of workshop:
Ranked by:
Date:

No.	Level (baseball rank) / Function	First Level (Little League)	Second Level (Junior Varsity)	Third Level (Varsity)	Fourth Level (Minor League Pro)	Fifth Level (Major League Pro)
1	Awareness Revolution	Large-scale mass production for maximum output	Product-out orientation	Maket-in orientation, but not implemented in each workshop	Service orientation, with service-oriented workshops	Service orientation implemented at each process factorywide
2	The 5S's	Hard for anyone to tell what goes where and when	Hard for visitors to tell what goes where and when, but workers know	Factory uses outlining and classification for visual control	Good indicators and clean, neatly organized factory	Clean, neatly organized with mess-prevention measures in force
3	Flow Production	Job-shop layout, geared for large-lot production	Job-shop layout, geared for small-lot production	In-line layout, small-lot flow at and between processes	In-line layout, one-piece flow at and between processes	Full multi-process operations with one-piece flow
4	Multi-process Operations	Unquestioned support for single-skill, single-process operations	Caravan-style cooperative operations	Flow-based cooperative operations	About halfway toward achieving smooth multi-process operations	Smooth and complete multi-process operations
5	Labor Cost Reduction	Wasteful motion and too many workers	Fixed job assignments and poor balance	Fixed job assignments, but different for each model, slightly better balance	About halfway toward achieving smooth multi-process operations	Flexible job assignments, with narrow variation in output volume
6	Kanban	Push production, with retained inventory all over the place	Push production, with organized storage sites for in-process inventory	Pull production, with fixed locations and fixed volumes	Flexible job assignments, with wide variation in output volume	Kanban and improvements
7	Visual Control	Abnormalities often occur and only create confusion	Abnormalities often occur and are usually resolved in some way	Supervisors can tell when an abnormality occurs	Pull production, with kanban	Immediate action is taken to resolve abnormalities
8	Level Production	Once-a-month production schedule, processes have own rhythms	Twice-a-month production schedule, each process has its own rhythm	Weekly production schedule, overall line has some kind of common rhythm	Anyone can tell when an abnormality occurs	Completely level production, overall line has a common rhythm
9	Changeover	Monthly changeover, requires half a day each time	People are aware of changeover needs	Changeover teams and improvements made in some workshops	Daily production schedule, overall line has a common rhythm Single-operation changeovers	Changeovers are within cycle times
10	Quality Assurance	Factory ships defective products and deals with customer complaints	Defective products are sorted out at final inspection and not shipped	Factory produces defective products, but passes information to reduce defects	Processes do not send defectives downstream (independent inspection)	Factory builds quality in at each process (at-the-source inspection)
11	Standard Operations	Operation procedures are generally left up to each operator	Operation procedures are vaguely standardized in roughly the same order	Standard operations implemented for individual processes	Standard operations planned, but not fully implemented	Standard operations and improvements fully implemented
12	Human Automation	All processes require manual assistance, lots of large-lot equipment	Some automation, but operators are always present while machines work	Human and machine work separated; machines sometimes make defective items	Human and machine work separate, but machines sometimes make defectives	Human and machine work separate, with no defectives, and with some human automation devices
13	Maintenance and Safety	Lots of breakdowns and numerous accidents per year	Factory uses maintenance specialists, but has occasional accidents	Factory has follow-up maintenance and no major accidents	Factory has preventive maintenance and is almost accident-free	Factory has company-wide preventive maintenance and no accidents

Figure 16.10   List of JIT's Major Functions and Their Five Stages of Development.

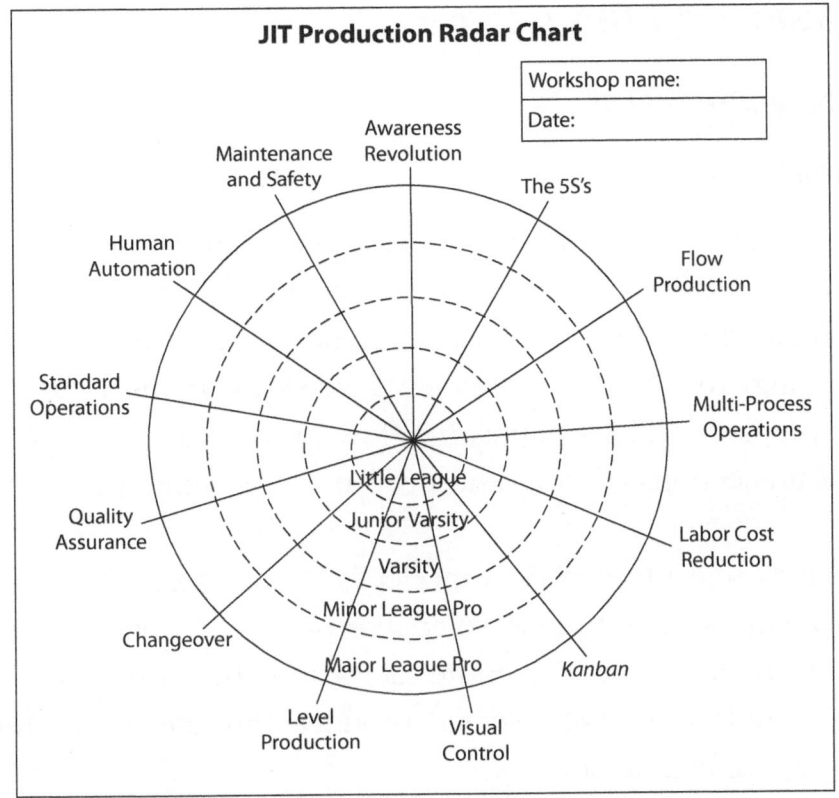

Figure 16.11   JIT Production Radar Chart.

Waste-Related Forms

Arrow Diagrams

Application

Arrow diagrams are useful for analyzing factory conditions in terms of the flow of goods to discover major forms of waste, such as conveyance waste and retention waste (see Figures 16.12 and 16.13). We should create arrow diagrams as part of our preparation for establishing flow production.

Main sections of form and procedure for filling out form:

1. First select the product model to be analyzed. Use a P-Q analysis sheet to help select the product model.
2. Draw a map of the entire factory or the workshop that manufactures the selected product. Indicate the current equipment layout.
3. Use the following process analysis symbols and indicate the sequence in which goods flow through the factory or workshop.
   a. Processing: large circle
   b. Inspection: diamond
   c. Conveyance: small circle
   d. Retention: triangle

   Example: An "F" inside a large circle can indicate forklift conveyance.
4. At all conveyance points, indicate the conveyance distance and the type of conveyance device used. At all retention points, indicate the usual amount of in-process inventory being retained.
5. Show totals for the number of retention points, conveyance points, processes, and inspections. Also show the total amount of retained goods and the total conveyance distance.
6. Devise and implement means of removing major forms of waste, such as conveyance waste and retention waste.

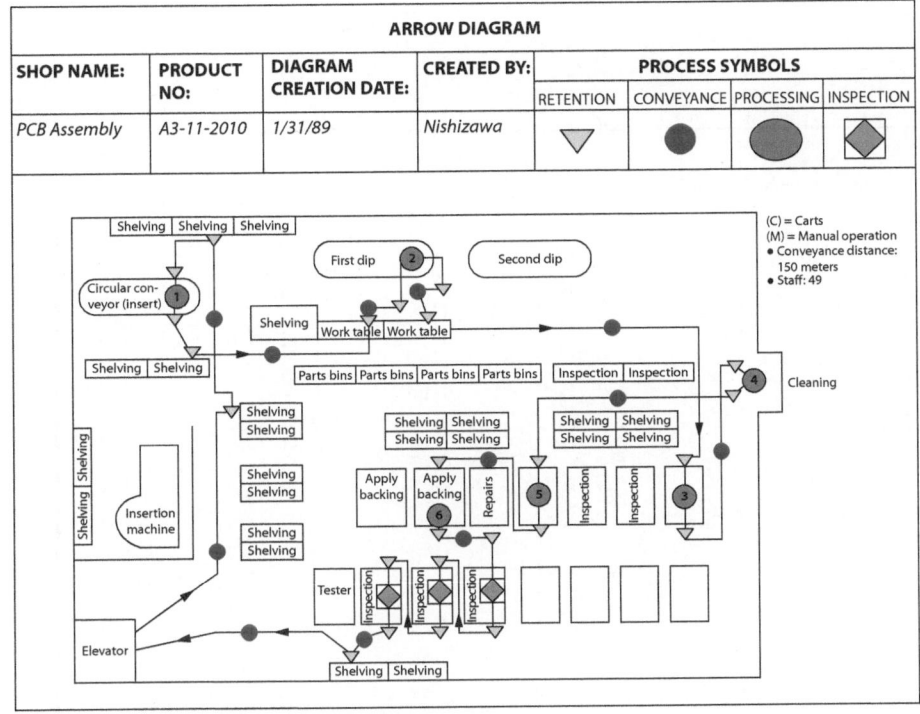

Figure 16.12   Example of an Arrow Diagram.

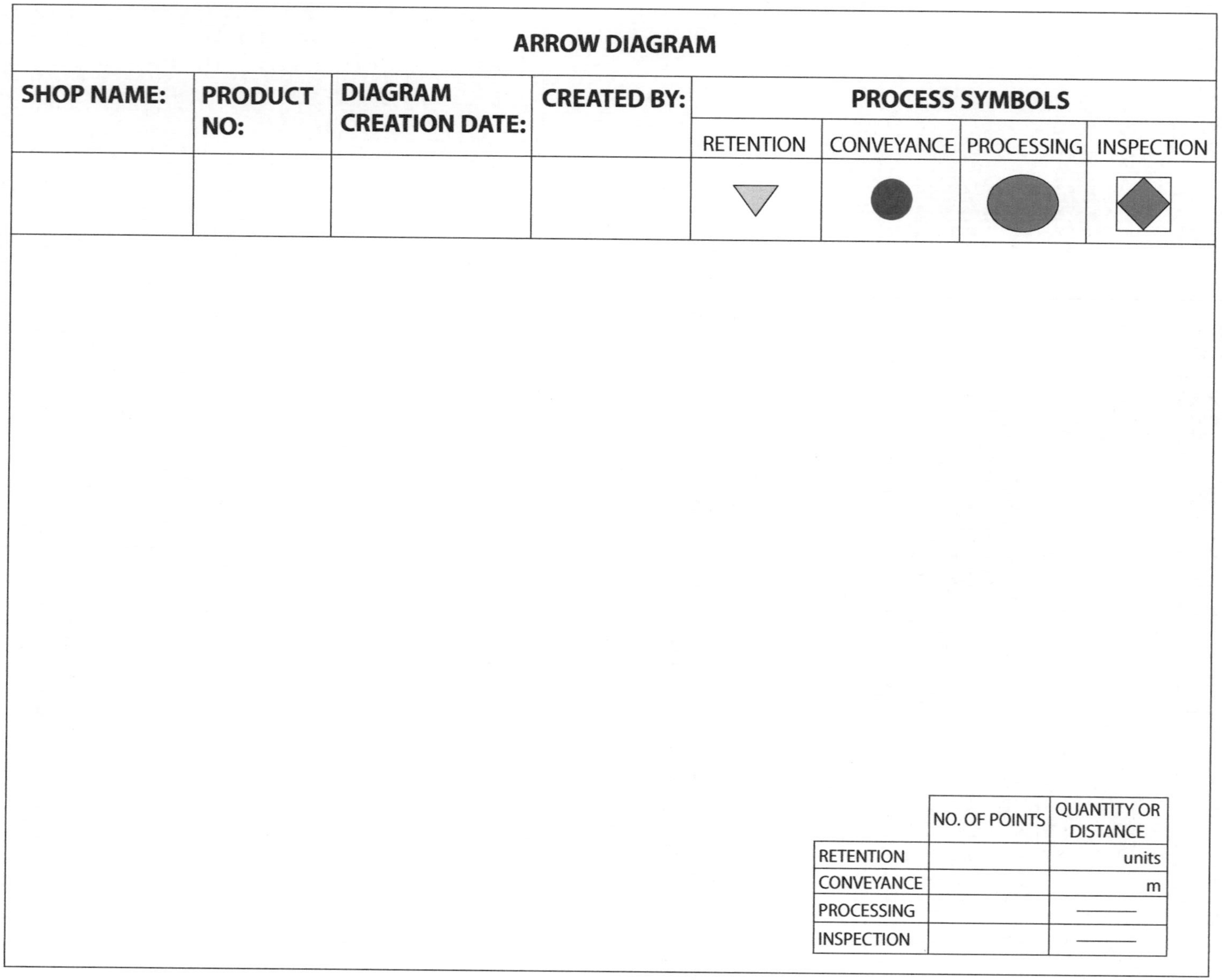

ARROW DIAGRAM

SHOP NAME:	PRODUCT NO:	DIAGRAM CREATION DATE:	CREATED BY:	PROCESS SYMBOLS		
				RETENTION	CONVEYANCE	PROCESSING	INSPECTION

NO. OF POINTS	QUANTITY OR DISTANCE
RETENTION	units
CONVEYANCE	m
PROCESSING	——
INSPECTION	——

Figure 16.13   Arrow Diagram.

General Flow Analysis Charts

Application

General flow analysis charts (see Figures 16.14 and 16.15) are useful for describing and comparing the production flow for a particular product before and after improvement.

Main sections of the form:

1. *Before improvement.* Analysis results describing retention, conveyance, processes, and inspection before improvement.
2. *After improvement.* Analysis results describing retention, conveyance, processes, and inspection after improvement.

Summary Chart of Flow Analysis

Date:

No.	Product name/No.: Part name/No.:	Before — Retention No. of times	Before — Retention No. of units	Before — Time	Before — Conveyance No. of times	Before — Distance	Before — Processing No. of times	Before — Lots	Before — Inspection No. of times	Before — Lots	After — Retention No. of times	After — Retention No. of units	After — Time	After — Conveyance No. of times	After — Distance	After — Processing No. of times	After — Lots	After — Inspection No. of times	After — Lots
1	PCB1 (A3-11-2010)	24	960	3	16	6	40	3	40	12	240	1	8	400	400	6	20	1	20
2	PCB2 (A6-63-1131)	20	800	3	1000	6	40	3	40	10	200	1	6	350	350	6	20	1	20
3	PCB3 (A4-21-3613)	22	880	3	1200	6	40	3	40	11	220	1	7	400	400	6	20	1	20
4	PCB4 (A3-16-2131)	24	960	3	1200	6	40	3	40	12	400	1	8	400	400	6	20	1	20
5	PCB5 (A6-23-61)																		

Figure 16.14   Example of a General Flow Analysis Chart.

Summary Chart of Flow Analysis

Date:

No.	Product name/No.: / Part name/No.:	Before improvement								After improvement								
		Retention			Conveyance		Processing		Inspection		Retention			Conveyance		Processing		Inspection
		No. of times	No. of units	Time	No. of times	Distance	No. of times	Lots	No. of times	Lots	No. of times	No. of units	Time	No. of times	Distance	No. of times	Lots	No. of times	Lots

Figure 16.15   General Flow Analysis Chart.

Operations Analysis Charts

Application

Use these charts (see Figures 16.16 and 16.17) to analyze the series of operations that go into manufacturing a particular product to help clarify and remove inherent waste, such as idle time waste, transfer waste, and unnecessary movement.

Main sections of form:

1. *Work.* Indicate each operation (processing) that directly adds value to the product, and thereby generates profit.
2. *Motion.* Indicate as motion anything that does not add value to the product, but supports processing that does add value. No matter how much of this type of "motion" we use, we are not generating any profit.
3. *Idle time.* Indicate idle time during manufacturing operations.
4. *Time.* Enter time measurements for operations (both work and motion), transfer, idle time, and inspection.
5. *Distance.* Enter all transfer distances.

Figure 16.16   Example of an Operations Analysis Table.

Operations Analysis Table					Section		Operation			
					Processes		Part No.		Author:	

	Before Improvement (date:)							After Improvement (date:)						
No.	Work	Movement	Transfer	Idle	Inspect	Description of operation	Time	Distance	Work	Movement	Transfer	Idle	Inspect	Description of operation	Time	Distance
	○ ● ● ▽ ◇								○ ● ● ▽ ◇						
1	○ ● ● ▽ ◇								○ ● ● ▽ ◇						
2	○ ● ● ▽ ◇								○ ● ● ▽ ◇						
3	○ ● ● ▽ ◇								○ ● ● ▽ ◇						
4	○ ● ● ▽ ◇								○ ● ● ▽ ◇						
5	○ ● ● ▽ ◇								○ ● ● ▽ ◇						
6	○ ● ● ▽ ◇								○ ● ● ▽ ◇						
7	○ ● ● ▽ ◇								○ ● ● ▽ ◇						
8	○ ● ● ▽ ◇								○ ● ● ▽ ◇						
9	○ ● ● ▽ ◇								○ ● ● ▽ ◇						
10	○ ● ● ▽ ◇								○ ● ● ▽ ◇						
11	○ ● ● ▽ ◇								○ ● ● ▽ ◇						
12	○ ● ● ▽ ◇								○ ● ● ▽ ◇						
13	○ ● ● ▽ ◇								○ ● ● ▽ ◇						
14	○ ● ● ▽ ◇								○ ● ● ▽ ◇						
15	○ ● ● ▽ ◇								○ ● ● ▽ ◇						
16	○ ● ● ▽ ◇								○ ● ● ▽ ◇						
	○ ● ● ▽ ◇								○ ● ● ▽ ◇						
	○ ● ● ▽ ◇								○ ● ● ▽ ◇						
	○ ● ● ▽ ◇								○ ● ● ▽ ◇						
	○ ● ● ▽ ◇								○ ● ● ▽ ◇						

Figure 16.17   Operations Analysis Table.

Waste-Finding Checklist
(Workshop-Specific and Process-Specific)

Application

This list helps us be thorough in finding how the seven major types of waste exist in each process or workshop. Making such a list is also an effective preparatory step before making improvements to establish flow production (see Figures 16.18 to 16.23).

Main sections of form:

1. Waste-finding Checklist (workshop-specific)
   a. *Process name.* Name of the process where waste is being found.
   b. *Major waste.* Enter the magnitude of waste (on a scale of 0 to 4) under each column.
      i. 0: No waste
      ii. 1: A little waste
      iii. 2: Some obvious waste
      iv. 3: Considerable waste
      v. 4: Lots of waste
   c. *Total waste magnitude.* Add up the magnitude points for all seven major types of waste.
   d. *Order of improvements.* Begin making improvements at the processes having the greatest magnitude of waste.
2. Waste-finding Checklist (process-specific)
   a. *Respond YES or NO to the statements in the form.* Example: "No production schedule or control boards."
      i. Answer YES if there are none.
      ii. Answer NO if there are some.
   b. *Magnitude of waste (1–3).*
      i. 1: A little waste
      ii. 2: Moderate waste
      iii. 3: A lot of waste

Workshop name:	**Waste-finding Checklist (workshop-specific)**				Date:

No.	Process name	1 Overproduction waste	2 Inventory waste	3 Conveyance waste	4 Defect-production waste	5 Processing-related waste	6 Operation-related waste	7 Idle time waste	Waste magnitude total	Improvement ranking	Improvement ideas and comments

Figure 16.18   Example of a Workshop-Specific Waste-Finding Checklist.

Waste-finding Checklist (process-specific) — Date:

Process Name / Major waste

Type 3. Conveyance waste	Description of waste	Confirmation YES	NO	Magnitude	Causes and improvement plans
	4. Pile-up during conveyance				
	5. Change of conveyance devices in mid-transfer				
	6. Previous and/ or next process is an another floor				
	7. Conveyance requires manual assistance				
	8. Conveyance distance too long				
	9.				
	10.				

Type 4. Defect-production waste
1. Complaints from r
2. Defect within proc
3. Human errors
4. Defective due to
5. Defective due to v
6. Omission in proce
7. Defect in processi
8. No human autom
9. No *poka-yoke*
10. No inspection wi
11. Defects not addr
12.

Type 5. Processing-related waste
1. Process is not requ
2. Process includes u
3. Process can be rep less wasteful
4. Part of process car detracting from p
5.

Type 6. Motion-related waste
1. Walking
2. Turning around
3. Leaning sideways
4. Leaning over
5. Wide-arm movem

Waste-finding Checklist (process-specific) — Date:

Process name / Major waste

Type 1. Overproduction
1. No production schedule or control boards
2. No leveling of production schedule
3. Production not in sync with production schedule
4. Items missing
5. Defective goods prod
6. Equipment breakdow
7. Too much manual ass
8. Too much capacity
9. Lots grouped into bat
10. Using push producti
11. Caravan-style operat
12. Not balanced with n

Type 2. Inventory waste
1. Lots of inventory on s
2. Shelves and floor stor
3. Inventory stacks block
4. In-process inventory a operations
5. In-process inventory i operators
6. In-process inventory i processes
7. Impossible to visually in-process inventory
8.
9.
10.
11.

Waste-finding Checklist (process-specific) — Date:

Process Name / Major waste

	Description of waste	Confirmation YES	NO	Magnitude	Causes and improvement plans
Type 6. Operation-related waste	6. Wrist movements				
	7. Left or right hand is idle				
	8. Idle time used for observation				
	9. Workpiece setup/removal				
	10. No standardized repetition of operations				
	11. Worker operates using different motions each time				
	12. Operations divided up into little segments				
	13.				
	14.				
	15.				
Type 7. Idle time waste	1. Idle time due to workpiece delay from previous piece				
	2. Idle time due to machine busy status				
	3. Idle time due to missing item(s)				
	4. Idle time due to lack of balance with previous process				
	5. Idle time due to lack of planning				
	6. Idle time due to lack of standard operations				
	7. Idle time due to worker absence				
	8. Idle time due to too many workers (more than two)				
	9.				
	10.				
	Total				

Overall improvement points:

Figure 16.19 Examples of Process-Specific Waste-Finding Checklists.

No.	Process name	1 Overproduction waste	2 Inventory waste	3 Conveyance waste	4 Defect-production waste	5 Processing-related waste	6 Operation-related waste	7 Idle time waste	Waste magnitude total	Improvement ranking	Improvement ideas and comments

Workshop name:

Waste-finding Checklist (workshop-specific)

Date:

Figure 16.20   Waste-Finding Checklist (Workshop-Specific).

Process name	**Waste-finding Checklist (process-specific)**				Date:	
Major waste	**Description of waste**	**Confirmation**		Mag-nitude	**Causes and improvement plans**	
		YES	**NO**			
Type 1. Overproduction	1. No production schedule or control boards				
	2. No leveling of production schedule				
	3. Production not in sync with production schedule				
	4. Items missing				
	5. Defective goods produced				
	6. Equipment breakdowns				
	7. Too much manual assistance				
	8. Too much capacity				
	9. Lots grouped into batches				
	10. Using "push production"				
	11. Caravan-style operations				
	12. Not balanced with next process				
Type 2. Inventory waste	1. Lots of inventory on shelves and floors				
	2. Shelves and floor storage take up lots of space				
	3. Inventory stacks block walkways				
	4. In-process inventory accumulates within individual operations				
	5. In-process inventory is stacked up between operators				
	6. In-process inventory is stacked up between processes				
	7. Impossible to visually determine quantities of in-process inventory				
	8.				
	9.				
	10.				
	11.				

Figure 16.21   Waste-Finding Checklist (Process-Specific) (1).

Process Name	Waste-finding Checklist (process-specific)				Date:
Major waste	Description of waste	Confirmation		Mag-nitude	Causes and improvement plans
		YES	NO		
Type 3. Conveyance waste	4. Pile-up during conveyance				
	5. Change of conveyance devices in mid-transfer				
	6. Previous and/or next process is on another floor				
	7. Conveyance requires manual assistance				
	8. Conveyance distance too long				
	9.				
	10.				
Type 4. Defect-production waste	1. Complaints from next process				
	2. Defect within process				
	3. Human errors				
	4. Defective due to missing part(s)				
	5. Defective due to wrong part(s)				
	6. Omission in processing				
	7. Defect in processing				
	8. No human automotion				
	9. No *poka-yoke*				
	10. No inspection with process				
	11. Defects not addressed by improvement activities				
	12.				
Type 5. Processing-related waste	1. Process is not required for product function				
	2. Process includes unnecessary operations				
	3. Process can be replaced by something less wasteful				
	4. Part of process can be eliminated without detracting from product				
	5.				
Type 6. Motion-related waste	1. Walking				
	2. Turning around				
	3. Leaning sideways				
	4. Leaning over				
	5. Wide-arm movements				

Figure 16.22   Waste-Finding Checklist (Process-Specific) (2).

Process Name	Waste-finding Checklist (process-specific) Date:				
Major waste	**Description of waste**	**Confirmation**		**Mag-nitude**	**Causes and improvement plans**
		YES	**NO**		
Type 6. Operation-related waste	6. Wrist movements				
	7. Left or right hand is idle				
	8. Idle time used for observation				
	9. Workpiece set-up/removal				
	10. No standardized repetition of operations				
	11. Worker operates using different motions each time				
	12. Operations divided up into little segments				
	13.				
	14.				
	15.				
Type 7. Idle time waste	1. Idle time due to workpiece delay from previous piece				
	2. Idle time due to machine busy status				
	3. Idle time due to missing item(s)				
	4. Idle time due to lack of balance with previous process				
	5. Idle time due to lack of planning				
	6. Idle time due to lack of standard operations				
	7. Idle time due to worker absence				
	8. Idle time due to too many workers (more than two)				
	9.				
	10.				
	Total				
	Overall improvement points:				

Figure 16.23   Waste-Finding Checklist (Process-Specific) (3).

5W1H Sheet

Application

The 5W1H sheet (see Figures 16.24 and 16.25) is designed to guide our inquiry into the primary causes of problems and/or abnormalities.

Main sections of form and procedure for filling out form:

1. Take what appears to be the problem, and write it down in order to indicate a possible needed improvement. (Example: The factory has a stockpile of part A.)
2. Ask "why" regarding the problem just described. (Example: Why has the factory stockpiled part A?)
3. Follow the arrow down to the next row of boxes and describe the current conditions as an answer to the above question. (Example: The people who receive part A from the subcontractor store those parts here.)
4. Ask "why" again regarding the conditions just described. (Example: "Why do those people store those parts here?") Repeat this pattern until you have asked "why" at least five times.
5. When you reach the final "why," start thinking about "how," that is, how to make an improvement that will correct the root cause uncovered by the series of "why" questions.

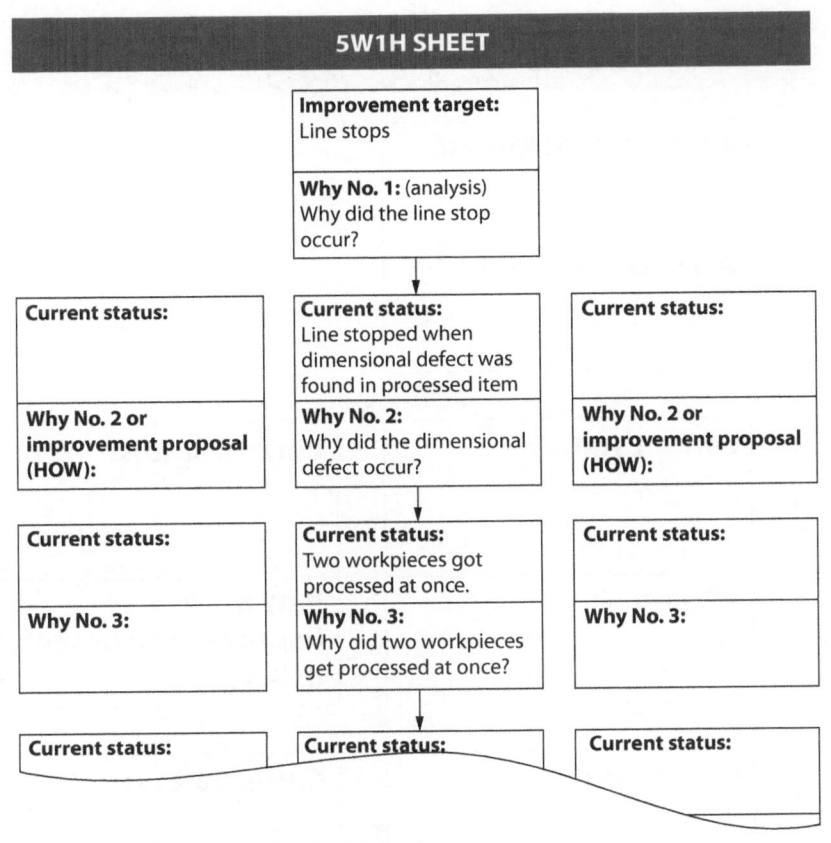

Figure 16.24 Example of a 5W1H Sheet.

5W1H SHEET

Improvement target:
Why No. 1: (analysis)

Current status:	Current status:	Current status:
Why No. 2 or improvement proposal (HOW):	**Why No. 2:**	**Why No. 2 or improvement proposal (HOW):**

Current status:	Current status:	Current status:
Why No. 3:	**Why No. 3:**	**Why No. 3:**

Current status:	Current status:	Current status:
Improvement proposal:	**Why No. 4:**	**Why No. 4:**

Current status:	Current status:	Current status:
Improvement proposal:	**Improvement proposal:**	**Improvement proposal:**

Figure 16.25   5W1H Sheet.

5S-Related Forms

5S Checklist (for Factories)

Application

We use this checklist (see Figures 16.26 and 16.27) to note how well the 5S's are being maintained in factories. We can also use this same checklist for checking 5S conditions at other factories, such as subcontractor factories.

Main sections of form:

1. *Location.* Distinguish between three types of general locations: outdoors, clerical, and factory.
2. *Year and month.* Enter which year and month during which the check-up is being done. If there are weekly checks, enter them in the columns below and add up a total figure under the "T" column.

Factory: Takai plant Checked by: Nishiro Kaibe		**5S Checklist (for factories)**	Scoring: 3 = Very good 2 = Good 1 = OK 0 = Not good						
			Year and month:						
Location	**Check item**	**Check description**	1	2	3	4	5	T	
Outdoors (overall)	Are there any unneeded items?	Outdoors (overall)	0	1	0	1	0	2	
	Are storage areas clearly determined?	Areas for: parking, pallets, temporary materials storage, delivered goods reception, trash processing, and boxes	0	2	0	2	0	4	
	Have paths been clearly defined?	Have white and yellow lines been laid down?	0	2	0	2	0	4	
		Are traffic signs being used?	0	3	0	3	0	6	
		Are there any exposed wires or pipes?	1	3	1	3	1	9	
	Are outdoor areas kept clean?	Are ashtrays, trash cans, gardens, entrance areas, windows, and paths kept clean?	1	3	1	3	1	9	
Clerical (overall)	Are there any unneeded items?	Are signboards, copy machines, and pathways arranged properly?	1	1	1	1	1	5	
	Have temporary storage areas been clearly defined?	Have fire-extinguishing equipment and emergency exits been established?	2	3	2	3	2	12	
	Are office areas being kept clean?	Are the walls dirty?	2	3	2	3	2	12	
		Is the area dusty?	2	3	2	3	2	12	
		Is the area decorated with fresh flowers?	2	2	2	2	2	10	
	Offices	Are there unneeded work utensils or other items on desks?	2	2	2	2	2	10	
		Are there unnecessary piles of paper on or in the desks?	1	2	1	2	1	7	
		Is anything being kept under the desk?	1	3	1	3	1	9	
		Is there a clearly defined area to store supplies?	1	1	1	1	1	5	
		Does the phone ring more than three times before being picked up?	1	1	1	1	1	5	
		Do people sit with straight backs?	0	1	0	1	0	2	
		in the lockers?	0	1	0	1	0	2	
					2	1	2	8	

Figure 16.26   Example of 5S Checklist.

Factory: Checked by:	**5S Checklist** **(for factories)**		Scoring: 3 = Very good 2 = Good 1 = OK 0 = Not good							
			Year and month:							
Location	**Check item**	**Check description**								
Outdoors (overall)	Are there any unneeded items?	Outdoors (overall)								
	Are storage areas clearly determined?	Areas for: parking, pallets, temporary materials storage, delivered goods reception, trash processing, and boxes								
	Have paths been clearly defined?	Have white and yellow lines been laid down?								
		Are traffic signs being used?								
		Are there any exposed wires or pipes?								
	Are outdoor areas kept clean?	Are ashtrays, trash cans, gardens, entrance areas, windows, and paths kept clean?								
Clerical (overall)	Are there any unneeded items?	Are signboards, copy machines, and pathways arranged properly?								
	Have temporary storage areas been clearly defined?	Have fire-extinguishing equipment and emergency exits been established?								
	Are office areas being kept clean?	Are the walls dirty?								
		Is the area dusty?								
		Is the area decorated with fresh flowers?								
	Offices	Are there unneeded work utensils or other items on desks?								
		Are there unnecessary piles of paper on or in the desks?								
		Is anything being kept under the desk?								
		Is there a clearly defined area to store supplies?								
		Does the phone ring more than three times before being picked up?								
		Do people sit with straight backs?								
		Are there unneeded articles in the lockers?								
		Do the lockers have designated compartments for necessary items?								
		Do trash cans or chairs block the walkways?								
		Do desks have ashtrays or are there ashes on the floor?								
	Meeting rooms	Are there any unnecessary items?								
		Do tables, desks, and chairs have designated locations?								
		Do desks have ashtrays or are there ashes on the floor? Are there any unneeded signs on the wall?								
		Are the names of meeting participants displayed somewhere?								
	Restrooms	Are there any unneeded items?								
		Are soap and paper towel dispensers kept stocked?								
		Are the floor and sink areas kept clean?								
		Is there any graffiti in the stalls?								

Figure 16.27   5S Checklist (for Factories).

5S Checklist (for Workshops) and 5S Radar Chart

Application

This is the workshop-specific version of the 5S checklist, which can be used to gather data and scores for "5S contests" among the workshops or different clerical sections (see Figures 16.28 to 16.31).

Main sections of form:

1. *The 5S's.* Questions are asked specifically about each of the 5S's: *seiri* (proper arrangement), *seiton* (orderliness), *seiso* (cleanliness), *seiketsu* (cleaned up), and *shitsuke* (discipline).
2. *Scoring.* Workshop conditions are scored on a scale of 0 to 5 as follows:
   a. 0: Very bad
   b. 1: Bad
   c. 2: OK
   d. 3: Good
   e. 4: Very good

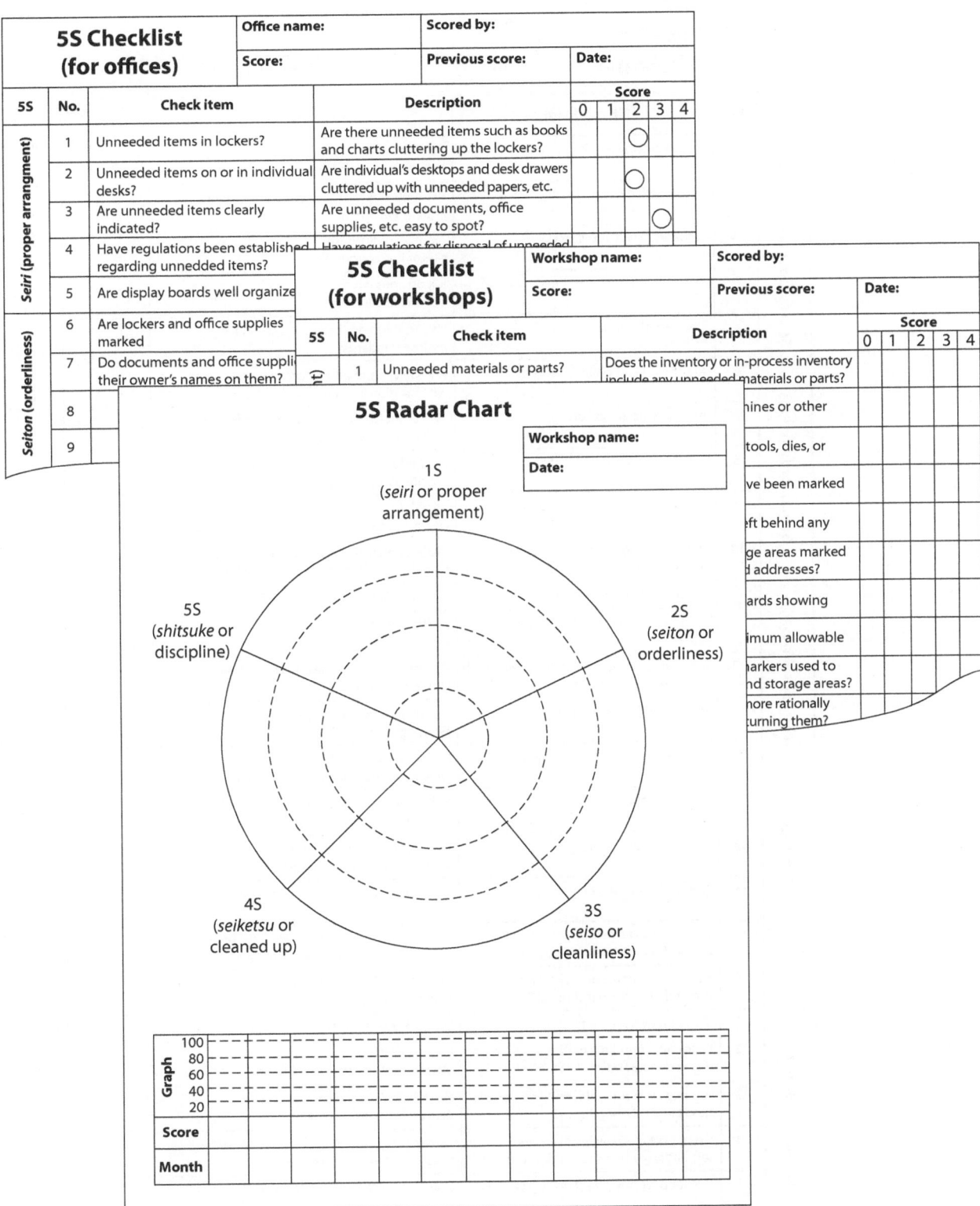

Figure 16.28   Sample 5S Chart.

5S Checklist (for workshops)		Workshop name:		Scored by:				
		Score:		Previous score:		Date:		

5S	No.	Check item	Description	Score				
				0	1	2	3	4
Seiri (proper arrangement)	1	Unneeded materials or parts?	Does the inventory or in-process inventory include any unneeded materials or parts?					
	2	Unneeded machines or other equipment?	Are there any unused machines or other equipment around?					
	3	Unneeded jigs, tools, or dies?	Are there any unused jigs, tools, dies, or similar items around?					
	4	Have unneeded items been marked?	Is it obvious which items have been marked as unnecessary?					
	5	Unneeded standards?	Has establishing the 5S's left behind any useless standards?					
Seiton (orderliness)	6	Are there location indicators?	Are shelves and other storage areas marked with location indicators and addresses?					
	7	Are there item indicators?	Do the shelves have signboards showing which items go where?					
	8	Are there quantity indicators?	Are the maximum and minimum allowable quantities indicated?					
	9	Demarcation of walkways and in-process inventory areas?	Are white lines or other markers used to clearly indicate walkways and storage areas?					
	10	Have improvements been made to facilitate jig and tool handling?	Are jigs and tools arranged more rationally to facilitate picking them up and returning them?					
Seiso (cleanliness)	11	Trash, water, or oil on floors?	Are floors kept shiny and clean?					
	12	Are machines covered with shavings and oil?	Are the machines wiped clean often?					
	13	Is equipment inspection combined with equipment maintenance?	Do operators clean their machines while checking them?					
	14	Have specific cleaning tasks been assigned?	Is there a person responsible for overseeing cleaning operations?					
	15	Has cleanliness become a habit?	Do operators habitually sweep floors and wipe equipment without being told?					
Seiketsu (cleaned up)	16	Is there proper ventilation?	Is the room ventilated well enough to be clear of heavy dust and odors?					
	17	Is there proper lighting?	Is the angle and intensity of the lighting adequate for the work being done?					
	18	Are work clothes clean?	Are workers wearing dirty or oil-stained work clothes?					
	19	Have improvements been made to prevent things from getting dirty?	Instead of cleaning up messes, have people found ways to avoid making messes?					
	20	Have rules been established for maintaining the first three S's?	Are the first three S's (*seiri, seiton,* and *seiso*) being maintained?					
Shitsuke (discipline)	21	Do workers have uniforms?	Do workers wear whatever they want?					
	22	Do people greet each other in the morning and say goodbye in the evening?	Do people verbally acknowledge each other when they happen to meet?					
	23	Are people punctual about their break times and meeting times?	Do people keep their appointments and take their breaks on time?					
	24	Do people casually review rules and regulations when they happen to meet?	Do people check with each other to confirm rules and correct procedures?					
	25	Do people obey rules and regulations?	Does each worker take rules and regulations seriously?					
Overall		Check for variation in scoring (note how many times)						

Figure 16.29   5S Checklist (for Workshops).

5S Checklist (for offices)		Offices name:		Scored by:					
		Score:		Previous score:		Date:			
5S	**No.**	**Check item**	**Description**	**Score**					
				0	1	2	3	4	

5S	No.	Check item	Description
Seiri (proper arrangement)	1	Unneeded items in lockers?	Are there unneeded items such as books and charts cluttering up the lockers?
	2	Unneeded items on or in individual desks?	Are individual's desktops and desk drawers cluttered up with unneeded papers, etc.
	3	Are unneeded items clearly indicated?	Are unneeded documents, office supplies, etc. easy to spot?
	4	Have regulations been established regarding unneeded items?	Have regulations for disposal of unneeded items been set?
	5	Are display boards well organized?	Are notices (concerning recreational activities, etc.) kept clean and displayed neatly?
Seiton (orderliness)	6	Are lockers and office supplies marked?	Are lockers and office supplies marked with location indicators?
	7	Do documents and office supplies have their owner's names on them?	Are such items marked with names to make identification easy?
	8	Are documents and office supplies easy to use?	Are documents and office supplies arranged so they are easy to pick up and put back?
	9	Are documents and office supplies kept where they are supposed to be kept?	Do documents and utensils have specified storage places and are they kept there?
	10	Are walkways and wall notices shown clearly?	
Seiso (cleanliness)	11	Trash or paper scraps on floors?	Are floors kept clean?
	12	Are windows and shelves dusty?	Are windows and shelves dusted and cleaned regularly?
	13	Have specific cleaning tasks been assigned?	Is there a person responsible for overseeing cleaning operations?
	14	Are trash cans allowed to overflow?	Do trash cans always get emptied before they overflow?
	15	Has cleanliness become a habit?	Do workers habitually sweep floors and wipe up dirt without being told?
Seiketsu (cleaned up)	16	Is there proper ventilation?	Is the room ventilated well enough to be clear of duct and cigarette smoke?
	17	Is there proper lighting?	Is the angle and intensity of the lighting adequate for the work being done?
	18	Are work clothes clean?	Are workers wearing dirty work clothes?
	19	Does the office give the impression of shining cleanliness at first sight?	Do the lighting, color design, and ventilation give a fresh atmosphere to the office?
	20	Have rules been established for maintaining the first three S's?	Are the first threes S's (*seiri, seiton,* and *seiso*) being maintained?
Shitsuke (discipline)	21	Do workers have uniforms?	Do workers wear whatever they want?
	22	Do people greet each other in the morning and say goodbye in the evening?	Do people verbally acknowledge each other when they happen to meet?
	23	Are people punctual about their break and meeting times?	Do people keep their appointments and take their breaks on time?
	24	Are people courteous and concise on the telephone?	Do they get to the point and make themselves understood?
	25	Do people obey rules and regulations?	Does each worker take rules and regulations seriously?
Overall		Check for variation in scoring (note how many times)	

Figure 16.30   5S Checklist (for Offices).

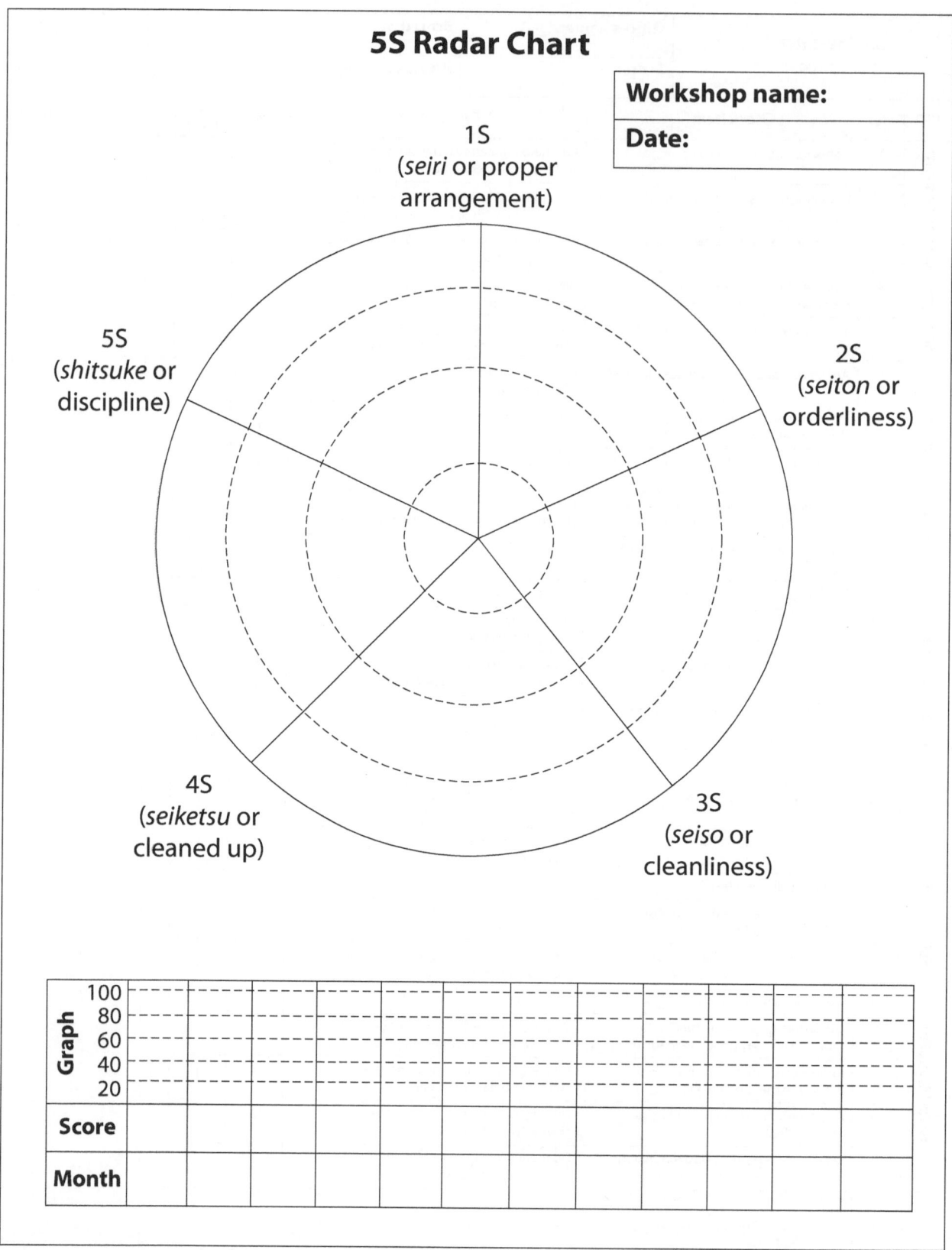

Figure 16.31   5S Radar Chart.

5S Memos

Application

5S memos are filled out by 5S patrols that tour the workshops and look for violations of 5S standards (see Figures 16.32 and 16.33). The patrols should take a photograph of the situation and attach a copy to the 5S memo describing the problem. After an improvement has been made to rectify the situation, an "after improvement" photo should also be taken for comparison and attached to the memo.

Main sections of form:

1. *Description of problem.* Point out the nature of the 5S violation.
2. *Implementation of improvement.* Describe the point of the improvement.

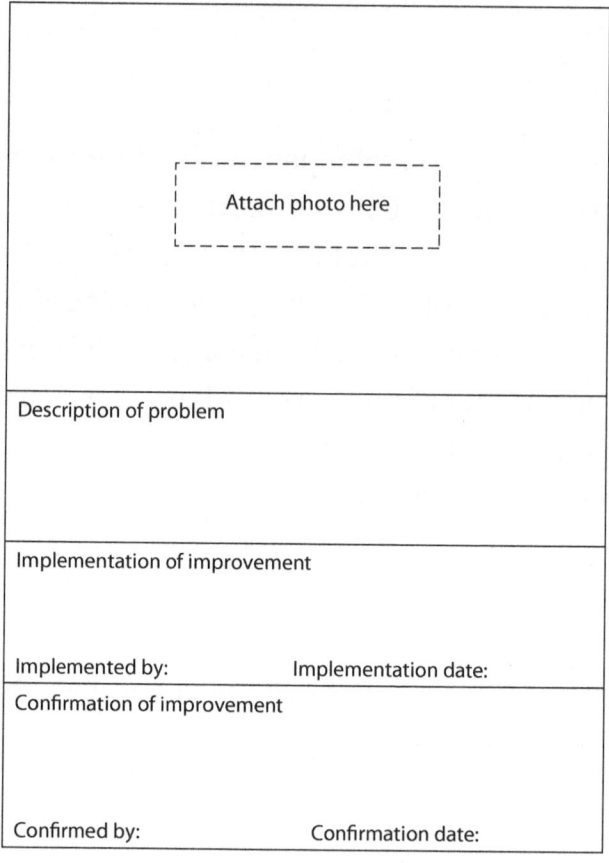

Figure 16.32   Example of 5S Memo.

Attach photo here	Attach photo here
Description of problem	Description of problem
Implementation of improvement Implemented by: Implementation date:	Implementation of improvement Implemented by: Implementation date:
Confirmation of improvement Confirmed by: Confirmation date:	Confirmation of improvement Confirmed by: Confirmation date:

Figure 16.33   5S Memo.

Red Tags

Application

Red tags (*Afakuda*), such as the one shown in Figures 16.34 and 16.35, are used in the 5S program's "red tag strategy." The point is to clearly mark unneeded items among the inventory, machines, and other tools and equipment.

Main sections of form:

1. *Category.* This is the category of the item marked with the red tag.
2. *Manufacturing number.* This number should be shown conspicuously and can be the lot number, manufacturing number, or other identifying number.
3. *Reason.* All red tags should identify a reason why the item was red-tagged.
4. *Disposal method.* The disposal method (or person to ask regarding disposal) should also be noted on each red tag.

RED TAG

Category	1. Raw material 2. In-process inventory 3. Semifinished product 4. Product	5. Manufacturing equipment 6. Die or jig 7. Tool or fixture 8. Other		
Item name				
Manufacturing No.				
Quantity	Units		Value	$
Reason	1. Not needed 2. Defective 3. Not needed soon 4. Scrap material 5. Use known	6. Other		
Disposal by:	**Department/Division/Section**			
Disposal method:	1. Discard 2. Return 3. Move to red tag storage site 4. Move to separate storage site 5. Other	**Disposal completed** (signature)		
Today's date:	**Posting date:**		**Disposal date:**	
Red tag file number				

Figure 16.34   Example of Red Tag.

RED TAG

Category	1. Equipment 2. Jigs and tools 3. Measuring instruments 4. Materials 5. Parts 6. In-process inventory	7. Quasi products 8. Finished products 9. Quasi materials 10. Office products 11. Paper, pens, etc.		
Item name				
Manufacturing No.				
Quantity	**Units**	**Value**		**$**
Reason	1. Not needed 6. Other 2. Defective 3. Not needed soon 4. Scrap material 5. Use known			
Disposal by:	**Department/Division/Section**			
Disposal method:	1. Discard 2. Return 3. Move to red tag storage site 4. Move to separate storage site 5. Other	**Disposal completed** (signature)		
Today's date:	**Posting date:**	**Disposal date:**		
Red tag file number				

Figure 16.35   **Red Tag.**

Red Tag Campaign Reports

Application

List the results of the red tag (*Afakuda*) campaign on this report under the categories of inventory, equipment dies/jigs/tools, and space considerations (see Figures 16.36 and 16.37).

Main sections of form:

1. *Target.* The type of item to which red tags were attached.
2. *Main items.* The main specific items to which red tags were attached.
3. *Number of red tags.* Indicate the total number of red tags used.
4. *Number of disposed items.* Indicate the number of items that have been disposed of so far.
5. *Value.* Indicate the cash value of the disposed items.

To:			Date:	
Red Tag Campaign Report				
Department: (signature of person responsible)				

Target		Main items	No. of red tags	No. of disposed items	Value
Inventory	Products				
	Parts				
	Materials				
Equipment					
Tools, jigs & dies					
Space		(Factory name)	(Space name)	(Person in charge)	(m²)
File locker					
Other					

Figure 16.36   Example of Red Tag Campaign Report.

To: Date:

Red Tag Campaign Report

Department: (signature of person responsible)

Target		Main items	No. of red tags	No. of disposed items	Value
Inventory	Products				
	Parts				
	Materials				
Equipment					
Tools, jigs & dies					
Space		(Factory name)	(Space name)	(Person in charge)	(m²)
File locker					
Other					

Figure 16.37   Red Tag Campaign Report.

Lists of Unneeded Inventory and Equipment

Application

All red-tagged inventory items should be listed on the "Unneeded Inventory List" and all red-tagged equipment units on the "Unneeded Equipment List." Note the disposal category and cash value for each item on either list (see Figures 16.38 to 16.40).

Main sections of form:

1. Unneeded Inventory List.
   a. *Code.* Item code.
   b. *Disposal category.* Indicate whether the item has been disposed of and, if so, by which method.
   c. *Supply category.* Indicate whether the item was supplied and, if so, whether it was paid for.
2. Unneeded Equipment List.
   a. *Asset Number.* Give the asset number from the listed equipment unit's procurement order.
   b. *Transaction date.* The date when the listed equipment unit was purchased.
   c. *Depreciation to date.* Indicate the total depreciation value to date.

Unneeded Inventory List

Division   Date:

Item	Code	Quantity	Unit value	Value	Disposal category	Supply category	Comments

Total value of unneeded items

Measures and improvement points:

Red Tag Campaign Report

To: Date:

Department: (signature of person responsible)

	Target	Main items	No. of red tags	No. of disposed items	Value
Inventory	Products				
Inventory	Parts				
Inventory	Materials				
	Equipment				
	Tools, jigs & dies				
	Space	(Factory name)	(Space name)	(Person in charge)	(m²)
	File locker				
	Other				

Figure 16.38 Example of Unneeded Inventory List and Red Tag Campaign Report.

Unneeded Inventory List

Division   Date:

Item	Code	Quantity	Unit value	Value	Disposal category	Supply category	Comments

| **Total value of unneeded items** | | | | | • Depreciation value
• Other | | |

Measures and improvement points:

Figure 16.39   Unneeded Inventory List.

						Depre-	Book		

Unneeded Equipment List   Date

Division:

Item	Asset No.	Quantity	Unit value	Transaction price	Transaction date	Depre- ciation to date	Book value	Location	Comments
Total value of unneeded items				—					
Response and improvement points									

Figure 16.40   Unneeded Equipment List.

Cleaning Checklist

Application

Workshop supervisors or cleanup inspection patrols can use this checklist to check how well cleaning tasks are being integrated with the daily maintenance tasks performed by equipment operators. As such, this checklist is a tool for company-wide maintenance activities (see Figures 16.41 and 16.42).

Main sections of form:

1. *Month.* Enter the current month (the year also, if necessary).
2. *Cleaning check points.* Draw a diagram of each workshop showing the equipment to be cleaned; have all important cleaning points marked with circled numbers. These will be the cleaning check points for cleaning inspectors.
3. *Date and day of week.* Note the date and the day of the week in which each inspection is done.
4. *Inspector.* Whoever is conducting the inspection should sign or initial this column.
5. *(1), (2), (3), etc.* The cleaning inspector checks off the box under each check point number as he or she inspects that check point and finds it clean.

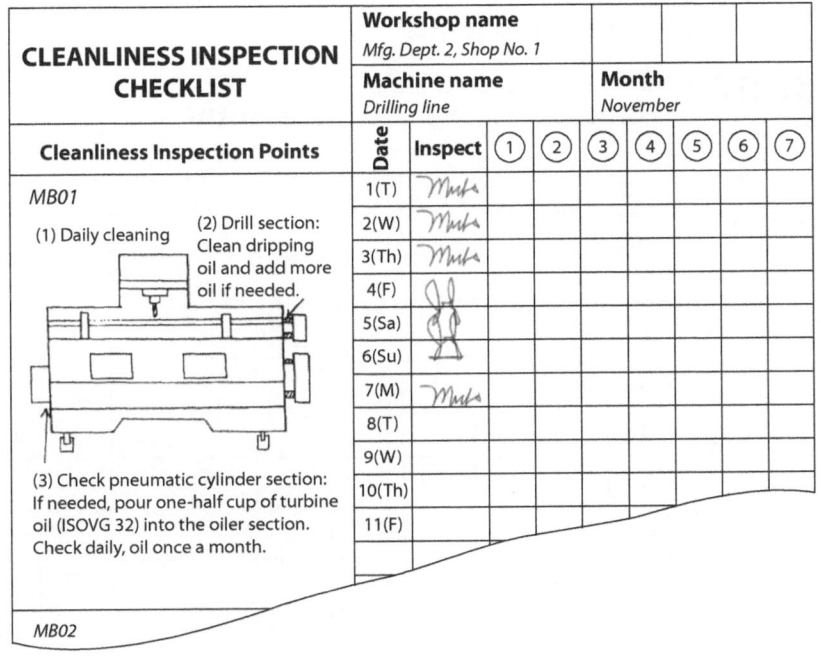

CLEANLINESS INSPECTION CHECKLIST		Workshop name *Mfg. Dept. 2, Shop No. 1*					
		Machine name *Drilling line*	Month *November*				
Cleanliness Inspection Points	Date Inspect	① ②	③ ④	⑤	⑥	⑦	

Figure 16.41 Example of Cleaning Checklist.

CLEANING CHECKLIST	Workshop name						
	Machine name		Month				
Cleaning Inspection Points	Date	Inspect ○	○	○	○	○	○ ○
	1 ( )						
	2 ( )						
	3 ( )						
	4 ( )						
	5 ( )						
	6 ( )						
	7 ( )						
	8 ( )						
	9 ( )						
	10 ( )						
	11 ( )						
	12 ( )						
	13 ( )						
	14 ( )						
	15 ( )						
	16 ( )						
	17 ( )						
	18 ( )						
	19 ( )						
	20 ( )						
	21 ( )						
	22 ( )						
	23 ( )						
	24 ( )						
	25 ( )						
	26 ( )						
	27 ( )						

Figure 16.42   Cleaning Checklist.

Five-Point Checklist to Assess Cleaned-Up Status

Application

To help maintain cleaned-up conditions, we can use this five-point checklist to rate the level of thoroughness in maintaining the first three "S's" (*seiri* or proper arrangement, *seiton* or orderliness, and *seiso* or cleanliness) for each workshop or process (see Figures 16.43 to 16.46).

(1) 5-point check for proper arrangement

Description Points	1	2	3	4	5
• Needed and unneeded items are mixed together in the workshop.	O				
• Needed and unneeded items are basically separated.		O			
• It is easy to see what is not needed.			O	O	O
• All unneeded items are kept somewhere away from the workshop.				O	O
• All completely unnecessary items have been disposed of.					O

(2) 5-point check for orderliness (warehouse and in-process inventory)

Description Points	1	2	3	4	5
• Can't tell what things belong where and in what amount.	O				
• Can basically tell what things belong where and in what amount.		O			
• Workshop is using only place indicators and item indicators.			O	O	O
• Workshop is using place and item indicators and outlining to make item organization visible.				O	O
• Input and output from workshop are clearly indicated and amount indicators are also being used.					O

(3) 5-point check for orderliness (for jigs and tools)

Description Points	1	2	3	4	5
• Can't tell what things belong where and in what amount.	O				
• Can basically tell what things belong where and in what amount.		O			
• Workshop is using only place indicators and jig/tool indicators.			O	O	O
• Measures have been taken to make item placement more visible (color coding, outlining, etc.)				O	O
• Jigs and tools have been streamlined by combining functions, etc.					O

Figure 16.43   5-Point Checks for Proper Arrangement.

(4) 5-point check for cleanliness

Description Points	1	2	3	4	5
• Workshop is left dirty for a long time.	O				
• Workers clean up the workshop occasionally.		O			
• Workers clean up the workshop daily.			O	O	O
• Daily cleaning tasks and maintenance have been integrated.				O	O
• Workers have devised ways to prevent messes.					O

Figure 16.44   5-Point Check for Cleanliness.

5-point "Cleaned-Up Checklist"		Factory name:	Division: *First assembly div.*	
		Date:	Entered by: *Yamaguchi*	Page: *1 of 1*

No.	Process and check point	Proper Arrangement	Orderliness	Cleanliness	Total	Previous total
1	Line A: operation at process 1	1 2 3 ④ 5	1 ② 3 4 5	1 ② 3 4 5	7	6
2	Line A: operation at process 2	1 ② 3 4 5	1 2 ③ 4 5	1 2 ③ 4 5	8	6
3	Line A: operation at process 3	1 ② 3 4 5	1 ② 3 4 5	1 ② 3 4 5	6	5
4	Line A: operation at process 4	1 ② 3 4 5	1 2 ③ 4 5	1 ② 3 4 5	7	7
5	Line A: operation at process 5	1 2 ③ 4 5	1 2 ③ 4 5	1 2 3 ④ 5	10	6
6	Line A: operation at process 6	1 2 3 ④ 5	1 2 3 ④ 5	1 2 3 ④ 5	12	8
	Line A: overall (average total)	②.8 1 2 3 4 5	②.8 1 2 3 4 5	②.8 1 2 3 4 5	㊼	㊳

Figure 16.45   Example of 5-Point Cleaned-Up Checklist.

	5-Point "Cleaned-Up Checklist"		Factory name:		Division:		
			Date:		Entered by:		Page:
No.	Process and check point	Proper Arrangement	Orderliness	Cleanliness	Total	Previous total
		1 2 3 4 5	1 2 3 4 5	1 2 3 4 5		
		1 2 3 4 5	1 2 3 4 5	1 2 3 4 5		
		1 2 3 4 5	1 2 3 4 5	1 2 3 4 5		
		1 2 3 4 5	1 2 3 4 5	1 2 3 4 5		
		1 2 3 4 5	1 2 3 4 5	1 2 3 4 5		
		1 2 3 4 5	1 2 3 4 5	1 2 3 4 5		
		1 2 3 4 5	1 2 3 4 5	1 2 3 4 5		
		1 2 3 4 5	1 2 3 4 5	1 2 3 4 5		
		1 2 3 4 5	1 2 3 4 5	1 2 3 4 5		
		1 2 3 4 5	1 2 3 4 5	1 2 3 4 5		
		1 2 3 4 5	1 2 3 4 5	1 2 3 4 5		
		1 2 3 4 5	1 2 3 4 5	1 2 3 4 5		
		1 2 3 4 5	1 2 3 4 5	1 2 3 4 5		
		1 2 3 4 5	1 2 3 4 5	1 2 3 4 5		

Figure 16.46   5-Point Cleaned-Up Checklist.

Display Boards

Application

Display boards provide clear indications of where things, such as in-process inventory and parts supplies, should be kept (see Figures 16.47 and 16.48).

Main sections of form:

1. *Temporary storage area.* The display board should indicate the location of the temporary storage site.
2. *Amount of stock.* Indicate both the maximum and the minimum amount to be stored at this site.
3. *Capacity.* Enter the capacity per container (bag, box, or whatever).
4. *Previous process and next process.* Write the names of the previous process (or company) and the next process (or company).
5. *Address.* If the storage site includes location indicators and addresses, such as is done for storage shelves, enter the address information here.

	A3-01-05		Temporary storage area
Amount of stock		Capacity	
Maximum *300* Minimum *500*	Bag Box Pallet	*50*	Bag Box Pallet
Previous process		Next process	
Person in charge		Address	
D. Lennon		Row Column Number	

Figure 16.47   Example of Display Board.

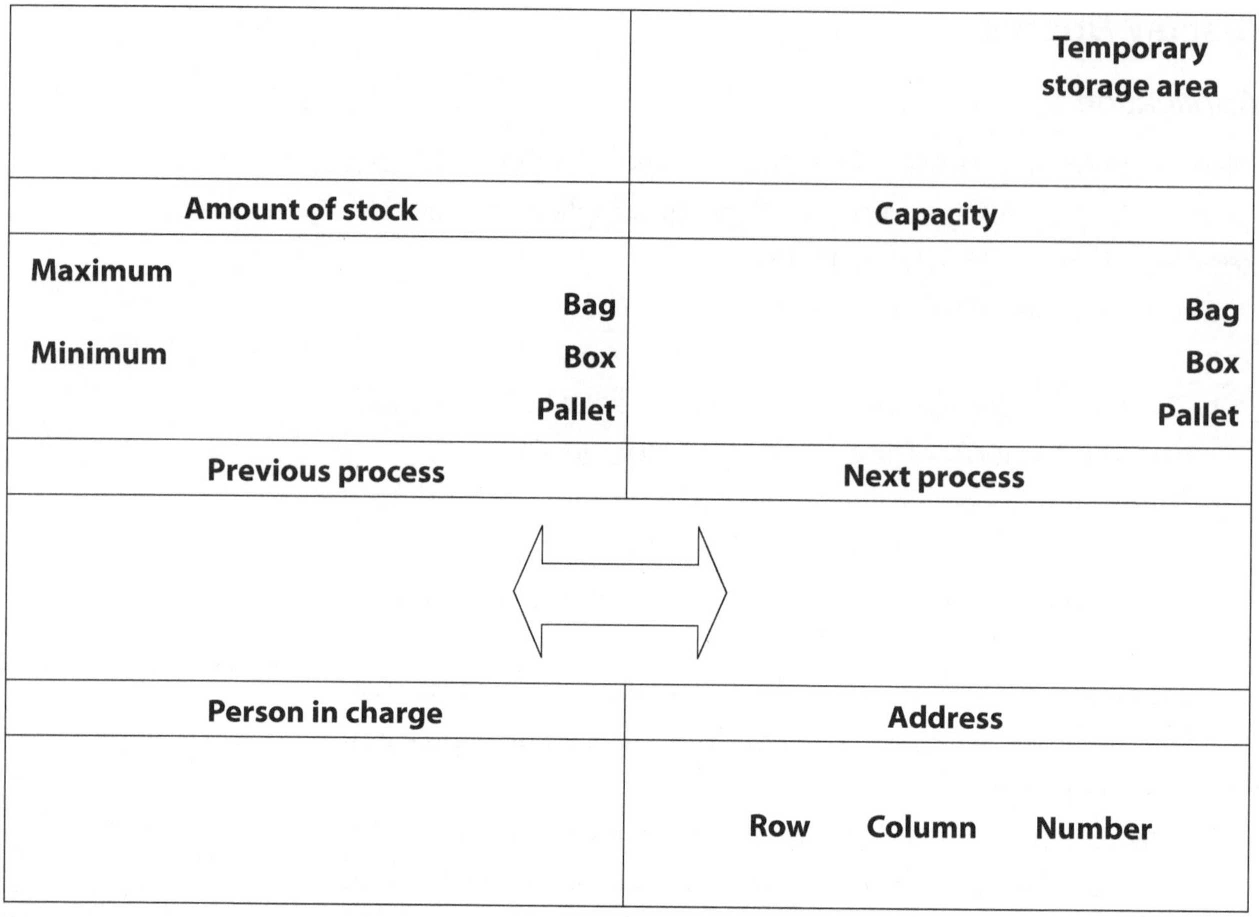

	Temporary storage area
Amount of stock	**Capacity**
Maximum **Bag** **Minimum** **Box** **Pallet**	**Bag** **Box** **Pallet**
Previous process	**Next process**
⇐⇒	
Person in charge	**Address**
	Row **Column** **Number**

Figure 16.48   **Display Board.**

Engineering-Related Forms

P-Q Analysis Lists and P-Q Analysis Charts

Application

We use product-quantity (P-Q) analysis lists and charts (see Figures 16.49 to 16.52) to estimate the output quantities of each type of product or finished component. This information can be useful when redesigning the equipment layout to facilitate flow production.

Main sections of form:

1. P-Q analysis list
   a. *Analysis period.* Enter the start and end dates of the analysis.
   b. *Item name (or number).* Enter the name or number of the item being analyzed. Start with items being output in the largest quantities.
   c. *Total.* Enter totals starting with the largest figures.
   d. *Percentage.* Give the percentage of that item within the total for all items.
   e. *Total percentage.* Give the percentage of that item and all previously listed items within the overall total quantity.
   f. *Management category.* Show which product category each item falls under.
2. P-Q analysis chart
   a. *Leftmost column.* Enter quantities in this column.
   b. *Part name.* Enter each part's quantity, with largest at left, to make a bar graph.
   c. *Percentage.* Enter the cumulative percentage points and connect the points to make a line graph.

No.	Item (part number)	Quantity	Total	%	Total %	Management category A	B	C
	P-Q Analysis List			Analysis by: *J. Smith*	Date: *11/16/89*			
				Analysis period: *10/1/89 to 10/31/89*				
1	RA1103	15,900	15,900	17.5	17.5	◯		
2	RB0121	12,500	28,400	13.7	31.2	◯		
3	RC1631	11,700	40,100	12.9	44.1	◯		
4	RD1911	9,450	49,550	10.4	54.5	◯		
5	RE0314	9,400	58,950	10.3	64.8	◯		
6	RF1213	9,000	67,950	9.9	79.7		◯	

Figure 16.49   Example of P-Q Analysis List.

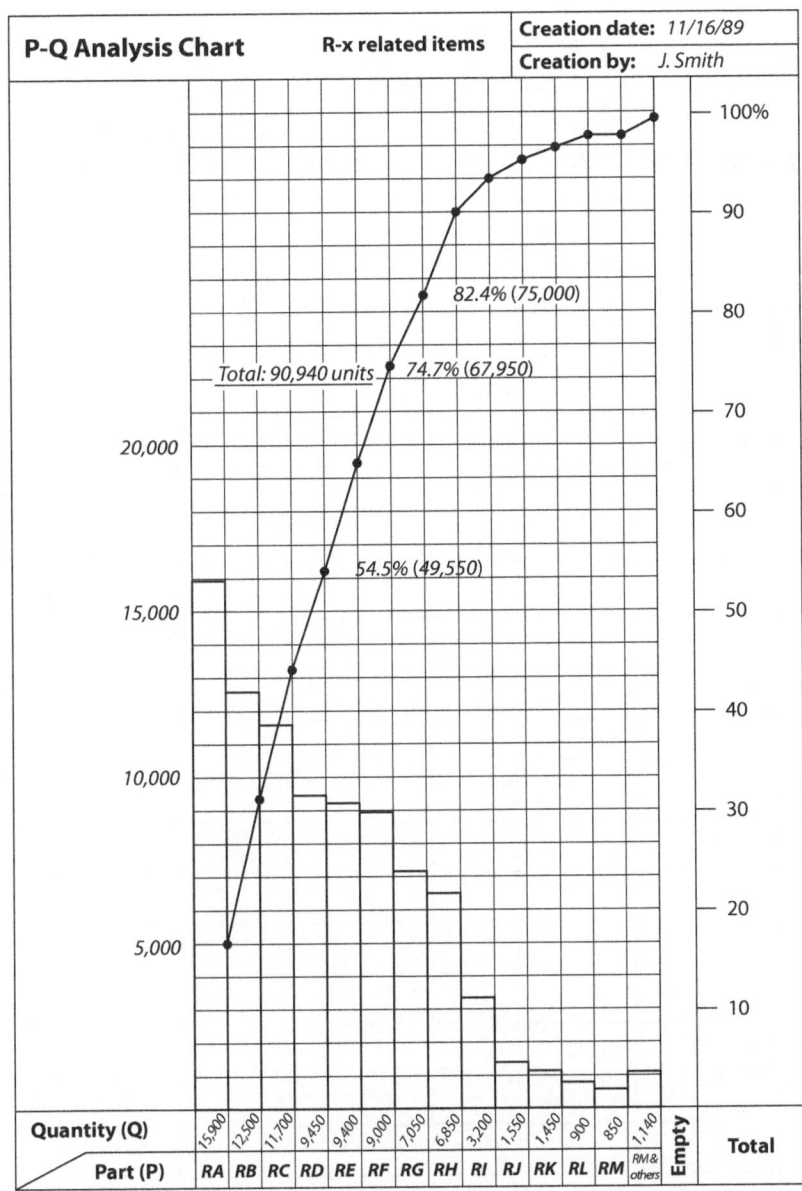

Figure 16.50 Example of P-Q Analysis Chart.

No.	Item (part number)	Quantity	Total	%	Total %	Management category		
						A	B	C

P-Q Analysis List

Analysis by: Date:

Analysis period:

Figure 16.51 P-Q Analysis List.

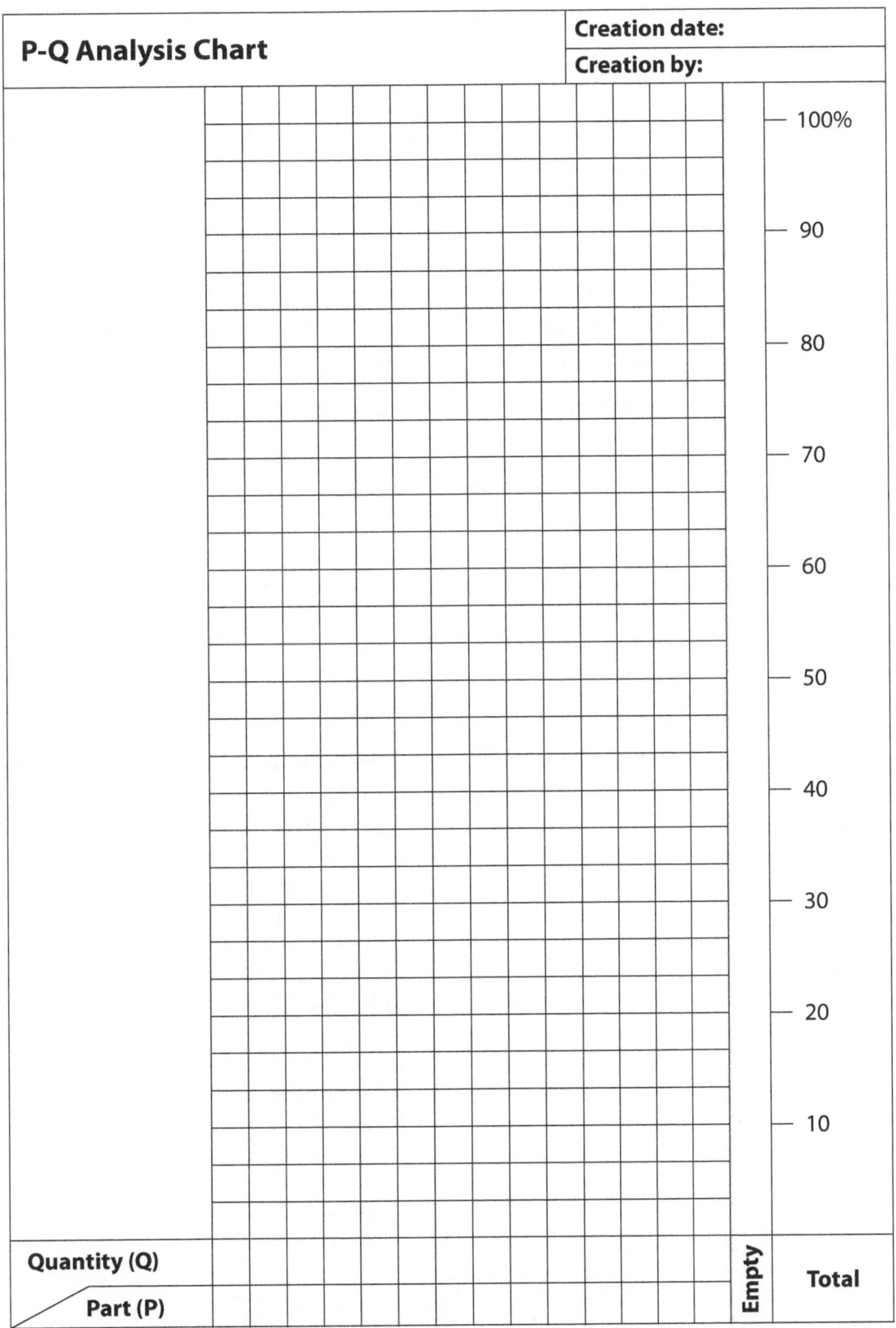

Figure 16.52 P-Q Analysis Chart.

Process Route Diagrams

Application

Process route diagrams illustrate the production flow for each type of workpiece and clarify the relationship between workpieces and machines (see Figures 16.53 and 16.54). They can therefore be used to get ideas for in-line equipment layout conducive to flow production. They are especially effective for processes that handle various types of workpieces and for designing G-T production lines.

Main sections of form and procedure for filling out form:

1. Enter the names of all of the processes and the numbers of the machines across the top of the columns.
2. Enter the names of the parts.
3. Enter circled numbers to indicate the machine numbers used for each part and draw lines to connect the circled numbers.
4. Find out if any of the parts use the same sequence of machines, and use those matching process routes when designing an in-line equipment layout.

No.	Process name / Machine no. / Item	Cutting M1	Drilling M2	Punching M3	Punching M4	Press M5	Press M6	Press M7	Bending M8	Bending M9	Bending M10			
1	110931 (side board)	①		②		③	④		⑤		⑥			
2	130106 (side board)	①		②		③		④		⑤				
3	161137 (side board)		①	③ ②			④	⑤			⑥			
4	1316171 (top board)		①	②		③			④					
5	1315021 (top board)		①		②			③		④				

Process Route Diagram — Factory: *Tokai Plant* — Entered by: *Shin'ichi Yamagawa* — Date: *January 10, 1989*

Figure 16.53   Example of Process Route Diagram.

Process Route Diagram	Factory:		Entered by:		Date:									
Item ╲ Process name ╲ Machine no. No.														

Figure 16.54   Process Route Diagram.

Line Balance Analysis Charts

Application

These charts help clarify the balance of work on the assembly line so that the balance can be improved to enable a smoother flow of goods with less line balance loss (see Figures 16.55 and 16.56).

Main sections of form:

1. *Top section.* Enter such factors as conveyor speed (COV. S), standard time (ST), and pitch time (PT).
2. *Process name.* Enter the process names in order from the left.
3. *Operation time analysis.* Clock the net operation time for each process and enter the figure in the "Time" column.
4. *Process time.* Plot the time for each process in the bar graph section.
5. *Line balance efficiency ratio and loss ratio.* Calculate and enter the line balance efficiency ratio and loss ratio.
6. *Identify bottleneck processes and improve line balance.* Use the chart to find out which processes have bottlenecks, analyze the operations at those processes, then move some of the work to the previous and/or next process to improve the line balance.
7. After the line balancing improvement has been made and tried out, conduct another line balance analysis to confirm the results.

Line Balance Analysis Chart

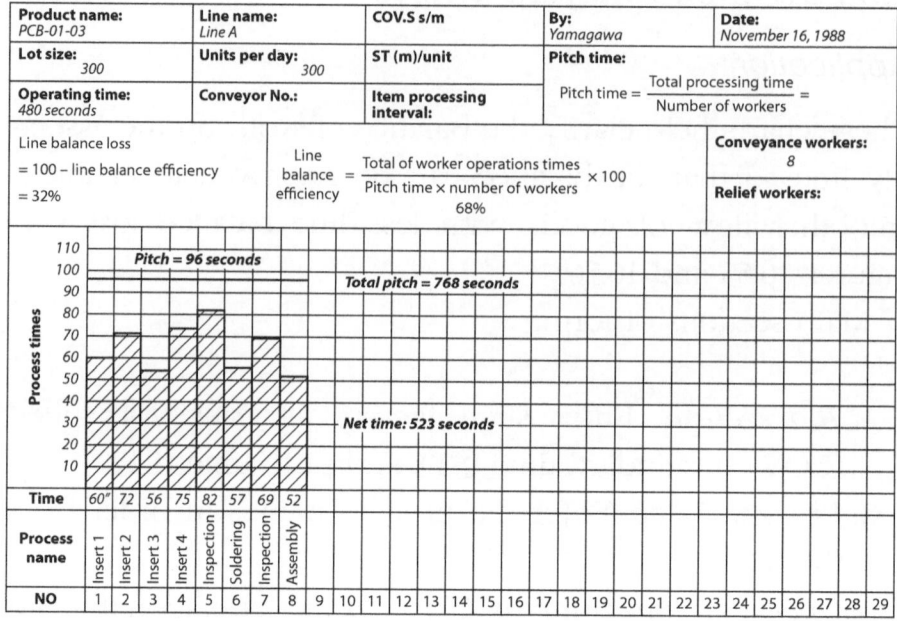

Product name: *PCB-01-03*	Line name: *Line A*	COV.S s/m	By: *Yamagawa*	Date: *November 16, 1988*
Lot size: *300*	Units per day: *300*	ST (m)/unit	Pitch time: Pitch time = $\dfrac{\text{Total processing time}}{\text{Number of workers}}$ =	
Operating time: *480 seconds*	Conveyor No.:	Item processing interval:		

Line balance loss
= 100 – line balance efficiency
= 32%

Line balance efficiency = $\dfrac{\text{Total of worker operations times}}{\text{Pitch time} \times \text{number of workers}} \times 100$
68%

Conveyance workers:
8

Relief workers:

Pitch = 96 seconds

Total pitch = 768 seconds

Net time: 523 seconds

Time	60"	72	56	75	82	57	69	52																					
Process name	Insert 1	Insert 2	Insert 3	Insert 4	Inspection	Soldering	Inspection	Assembly																					
NO	1	2	3	4	5	6	7	8	9	10	11	12	13	14	15	16	17	18	19	20	21	22	23	24	25	26	27	28	29

Figure 16.55   Example of Line Balance Analysis Chart.

Line Balance Analysis Chart

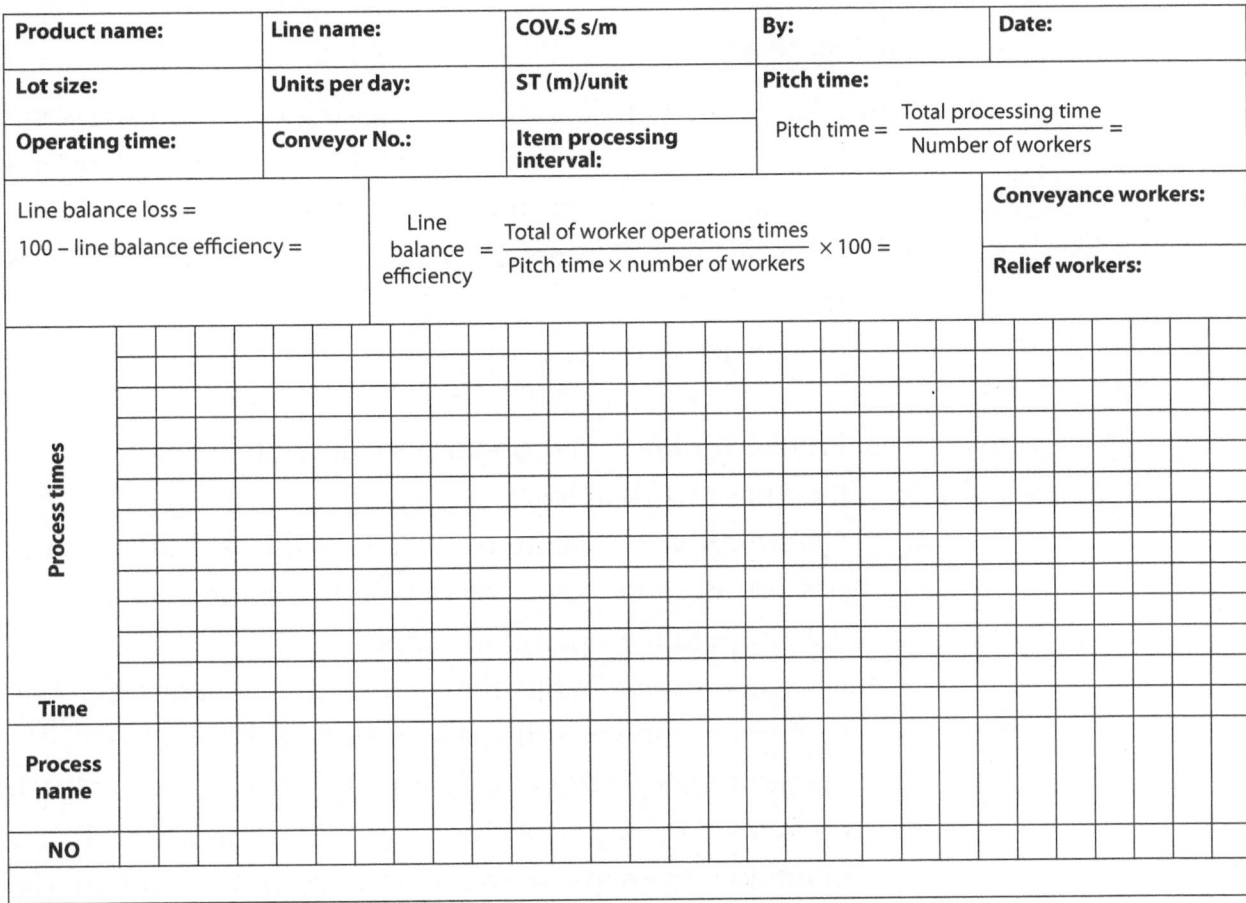

Product name:	Line name:	COV.S s/m	By:	Date:
Lot size:	Units per day:	ST (m)/unit	Pitch time:	
Operating time:	Conveyor No.:	Item processing interval:	Pitch time = Total processing time / Number of workers =	

Line balance loss = 100 − line balance efficiency =

Line balance efficiency = Total of worker operations times / Pitch time × number of workers × 100 =

Conveyance workers:

Relief workers:

Process times

Time

Process name

NO

Figure 16.56 Line Balance Analysis Chart.

Cooperative Operation Confirmation Chart

Application

This chart is used to confirm the passing of work on assembly lines in order to establish a baton touch zone for cooperative operations. As such, the chart helps eliminate omission of operations (see Figures 16.57 and 16.58) and other causes of defects.

Main sections of form:

1. *Process number.* The operators call out these numbers as they pass work along.
2. *Operation name.* Write the name of the operation being passed. Be sure to make detailed divisions of work in the cooperative operations zones.
3. *Operator name.* Write the names of the assembly line workers in the vertical rows at the left of the chart.
4. *Cooperative operation columns.* Enter a circle by each operator for each operation he or she handles. If the operation overlaps between two operators, fill in the circle. Make sure that operators practice the overlapped operations to become proficient.

Cooperative Operation Confirmation Chart		Factory: *Chiba*				Product: *PCB 1013*					
		Section: *1st Assembly Dpt., Line A*				By: *Yamagawa*			Date: *1/4/1989*		
Process No.		1	2	3	4	5	6	7	8		
No.	Operator name / Parts input	Tucker	Engle	North	Brown	Meyer	Kline	Jones	Black		
1	11-1640-20	○									
2	16-1311-31	○									
3	19-2931-16	●	●								
4	20-2131-16	●	●								
5	14-1923-61		○								
6	36-3111-21		○								
7	63-1416-41		●	●							
8	27-2131-51		●	●							
9	32-8136-24			○							

Figure 16.57   Example of Cooperative Operation Confirmation Chart.

Cooperative Operation Confirmation Chart		Factory:			Product:					
		Section:			By:			Date:		
Process No.										
No.	Parts input / Operator name									

Figure 16.58   Cooperative Operation Confirmation Chart.

Delivery Company Evaluation Charts

Application

We use this chart to evaluate the delivery methods (loading method, frequency of deliveries, transportation method, and so forth) used by parts suppliers and other subcontractors. As such, these charts (see Figures 16.59 and 16.60) provide valuable information that can be used in planning improvements.

Main sections of form:

1. *Main product.* Enter the main product that is being delivered by the company.
2. *Loading method.* Evaluate whether the current loading method is conducive to product diversification.
   a. 1 point: Lots containing identical items
   b. 2 points: Mixed lots (various items mixed together as a lot)
   c. 3 points: Sequential mixed lots (mixed lots loaded in the order the items will be used)
3. *Frequency of deliveries.* Evaluate whether their deliveries are frequent enough to keep inventory levels down and lead-times short.
   a. 1 point: Once a month or less
   b. 2 points: About once a week
   c. 3 points: About once every three days
   d. 4 points: Daily
   e. 5 points: About twice a day
   f. 6 points: About four times a day
   g. 7 points: At least eight times a day
4. *Transport routes.* Evaluate whether their transport routes are efficient enough to help hold down costs.
   a. 1 point: Point-to-point deliveries
   b. 2 points: Circuit deliveries
   c. 3 points: Compound deliveries
5. *Total points.* Add up and enter the points given in 2, 3, and 4 above.

No.	Company	Main product	Manager (in-house)	Manager (delivery company)	Loading method			Frequency of deliveries							Transport route			Total
					1	2	3	1	2	3	4	5	6	7	1	2	3	
1	M Company	Resistors	Off	Jones	○			○							○			3
2	Y Company	A1 units	Lennon	Sandler		○				○					○			6
3	K Company	C materials	Lennon	McTighe		○			○						○			5
4	F Company	Packaging	Off	Rosen		○						○			○			8
5	T Company	Coils	Smith	Amick	○				○						○			4

Delivery Company Evaluation Chart

Factory: Tohoku Plant

Name/Dept. of evaluator: Anderson, Purchasing dept.

Date: November 16, 1988

Figure 16.59   Example of Delivery Company Evaluation Chart.

No.	Company	Main product	Manager (in-house)	Manager (delivery company)	Loading method			Frequency of deliveries							Transport route			Total
					1	2	3	1	2	3	4	5	6	7	1	2	3	

Figure 16.60 Delivery Company Evaluation Chart.

JIT Delivery Efficiency List

Application

We can use these lists (see Figures 16.61 and 16.62) to rank parts suppliers and subcontractors on how well their deliveries live up to the Just-In-Time concept.

Main sections of form:

1. *Month.* Enter the month during which the evaluation is being made.
2. *Main product.* Enter the main product delivered by the company.
3. *Ranking.* List the companies in order of highest to lowest point-receivers.
4. *Volume of orders.* Enter the total amount of orders during the month under evaluation.
5. *Evaluation points.* Use the following formula to calculate the evaluation points for each company based on the points calculated for the JIT delivery efficiency column.

$$\text{Evaluation points} = J + (a + b + c + d)$$

6. *JIT delivery efficiency.* Calculate the percentage and the evaluation points as follows, based on the delivered amounts prior to JIT delivery time J (a & b) and after JIT delivery time J (c & d).
   a. After obtaining the values for a, b, c, and d, enter them in the header columns, as shown in Figure 16.61.

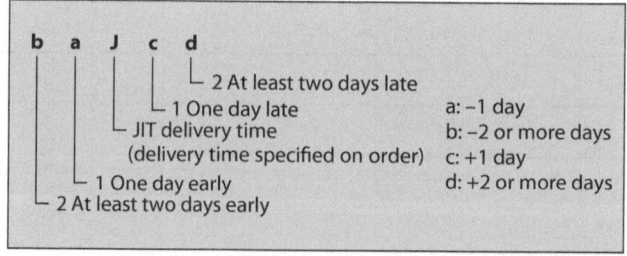

No.	Company	Main product	Rank	Order volume / Evaluation points	JIT delivery efficiency		a: b:	c: d:		Comments
					b (–2)	a (–1)	J	c (–3)	d (–4)	

JIT Delivery Efficiency List

Department
Month: By: Date:

No.	Order volume row	b (–2)	a (–1)	J	c (–3)	d (–4)	Comments
1		(%)	(%)	(%)	(%)	(%)	
2		(%)	(%)	(%)	(%)	(%)	
3		(%)	(%)	(%)	(%)	(%)	
		(%)	(%)	(%)	(%)	(%)	
		(%)	(%)	(%)	(%)	(%)	
		(%)	(%)	(%)	(%)	(%)	
		(%)	(%)	(%)	(%)	(%)	
		(%)	(%)	(%)	(%)	(%)	
		(%)	(%)	(%)	(%)	(%)	
		(%)	(%)	(%)	(%)	(%)	
		(%)	(%)	(%)	(%)	(%)	
		(%)	(%)	(%)	(%)	(%)	
		(%)	(%)	(%)	(%)	(%)	

Figure 16.61 Example of JIT Delivery Efficiency List.

b. Enter the amounts for J and a, b, c, and d, then calculate the percentage shares of each.

c. Assign "weight" values to a, b, c, and d based on their percentage shares, and enter these as evaluation points.

$$J = 1 \quad a = -1 \quad b = -2 \quad c = -3 \quad d = -4$$

No.	Company	Main product	Rank	Order volume / Evaluation points	JIT delivery efficiency					Comments
					b (−2)	a (−1)	J	c (−3)	d (−4)	a: c: b: d:
					(%)	(%)	(%)	(%)	(%)	
					(%)	(%)	(%)	(%)	(%)	
					(%)	(%)	(%)	(%)	(%)	
					(%)	(%)	(%)	(%)	(%)	
					(%)	(%)	(%)	(%)	(%)	
					(%)	(%)	(%)	(%)	(%)	
					(%)	(%)	(%)	(%)	(%)	
					(%)	(%)	(%)	(%)	(%)	
					(%)	(%)	(%)	(%)	(%)	
					(%)	(%)	(%)	(%)	(%)	
					(%)	(%)	(%)	(%)	(%)	
					(%)	(%)	(%)	(%)	(%)	

Department

Month: **By:** **Date:**

Figure 16.62 JIT Delivery Efficiency List.

Multiple Skills Training Schedule

Application

This schedule is a tool for training workers in the multiple skills needed for multi-process operations. It is the same as that of the "Multiple Skills Map." This training schedule is also known as the "Multiple Skills Score Sheet" and can be used in competitive games among trainees (see Figures 16.63 and 16.64).

Main sections of form:

1. *Operator's name.* Enter the names of the operators being trained in multiple skills.
2. *Process name.* Enter the name of the target process for skills training.
3. *Form completion method.* Fill in the circles as shown below to indicate skill level. Enter the target dates for skill achievement as shown in Figure 16.63. For example, "8/20" means that the trainee should be able to handle the target process by August 20.

Figure 16.63 Example of Multiple Skills Training Schedule.

Multiple Skills Training Schedule	○ Unable to do operation (LOSS) ◑ Can generally do operation (TIE) ● Can do operation well (WIN)	Factory name:		Foreman:	
		By:		Date:	

No.	Operator name	Process no. / Process name																Current date	Target date
		○⁄	○⁄	○⁄	○⁄	○⁄	○⁄	○⁄	○⁄	○⁄	○⁄	○⁄	○⁄	○⁄	○⁄	○⁄	○⁄		
		○⁄	○⁄	○⁄	○⁄	○⁄	○⁄	○⁄	○⁄	○⁄	○⁄	○⁄	○⁄	○⁄	○⁄	○⁄	○⁄		
		○⁄	○⁄	○⁄	○⁄	○⁄	○⁄	○⁄	○⁄	○⁄	○⁄	○⁄	○⁄	○⁄	○⁄	○⁄	○⁄		
		○⁄	○⁄	○⁄	○⁄	○⁄	○⁄	○⁄	○⁄	○⁄	○⁄	○⁄	○⁄	○⁄	○⁄	○⁄	○⁄		
		○⁄	○⁄	○⁄	○⁄	○⁄	○⁄	○⁄	○⁄	○⁄	○⁄	○⁄	○⁄	○⁄	○⁄	○⁄	○⁄		
		○⁄	○⁄	○⁄	○⁄	○⁄	○⁄	○⁄	○⁄	○⁄	○⁄	○⁄	○⁄	○⁄	○⁄	○⁄	○⁄		
		○⁄	○⁄	○⁄	○⁄	○⁄	○⁄	○⁄	○⁄	○⁄	○⁄	○⁄	○⁄	○⁄	○⁄	○⁄	○⁄		
		○⁄	○⁄	○⁄	○⁄	○⁄	○⁄	○⁄	○⁄	○⁄	○⁄	○⁄	○⁄	○⁄	○⁄	○⁄	○⁄		

Figure 16.64   Multiple Skills Training Schedule.

a. Empty circle: No experience

b. Half-filled circle: Can handle about 80 percent of tasks

c. Filled-in circle: Can handle all tasks within the cycle time

Multiple Skills Achievement Chart

Application

This chart is another tool for promoting multiple skills training (see Figures 16.65 and 16.66). It helps us gain a grasp of each trainee's current level of skills achievement and can be used to gauge trainees' progress during skills training programs.

Main sections of form:

1. *Operator's name.* Enter the name of the multiple skills trainee.
2. *Operation.* Enter the name of the operation being taught to the trainee.
3. *Progress.* Indicate the degree of progress toward the current skill achievement goal.
4. *OK.* The chart should be OK'd by the workshop leader once a month.
5. *Form completion method.* Indicate the progress in the quarter-marked circles shown.
   a. Empty circle: Completely unable to do operation
   b. Quarter-filled circle: Able to do the operation if someone else does the set-up
   c. Half-filled circle: Can generally do the operation, but needs minor guidance
   d. Three-quarters filled circle: Can do the operation well except under unusual conditions
   e. Filled circle: Can do the entire operation well

We can also use different colors to distinguish between actual results and predictions, as shown:

   f. White: Not a target
   g. Black: Last year's results
   h. Shading: This year's predicted results
   i. Red: This year's actual results

Period: *Dec.–Jan. 1988*	Multiple Skills Score Sheet					Section chief's check					
	Manufacturing Dept. 1, Section 2					1	2	3	4	5	6
						7	8	9	10	11	12

Operator name \ Process name	Coater 1	Coater 2	DB	PL	MJ	BP	CD	Progress 50% 100%
Worker A	●	●	●	●	●	●	●	
Worker B	●	●	●	●	◑	⊕	●	
Worker C	⊕	⊕	⊕	●	●	◑	⊕	
Worker D	●	⊕	●	●	●	⊕	⊕	
Worker E	⊕	●	◔	⊕	⊕	⊕	⊕	
Worker F	●	⊕	⊕	⊕	⊕	⊕	⊕	

Evaluation criteria

⊕ Unable to do operation

◔ Able to do the operation if someone else does the set-up

◔ Can generally do operation, needs minor guidance

◕ Can do the operation well, except under unusual circumstances

● Can do entire operation well

Color coding

Black .... 1987 results
Red shading .... Estimated 1988 results
Red .... 1988 results

Figure 16.65   Example of Multiple Skills Score Sheet.

Multiple Skills Achievement Chart	Evaluation criteria			By:					
	⊕ Unable to do operation	◔ Can do operation well, except under unusual circumstances	Date:						
	◐ Able to do operation if someone else does the set-up			1	2	3	4	5	6
	◕ Can generally do operation, needs minor guidance	● Can do entire operation well	OK	7	8	9	10	11	12

No.	Operator \ Operation											Progress 50%	Progress 100%
		⊕	⊕	⊕	⊕	⊕	⊕	⊕	⊕	⊕	⊕		
		⊕	⊕	⊕	⊕	⊕	⊕	⊕	⊕	⊕	⊕		
		⊕	⊕	⊕	⊕	⊕	⊕	⊕	⊕	⊕	⊕		
		⊕	⊕	⊕	⊕	⊕	⊕	⊕	⊕	⊕	⊕		
		⊕	⊕	⊕	⊕	⊕	⊕	⊕	⊕	⊕	⊕		
		⊕	⊕	⊕	⊕	⊕	⊕	⊕	⊕	⊕	⊕		
		⊕	⊕	⊕	⊕	⊕	⊕	⊕	⊕	⊕	⊕		
		⊕	⊕	⊕	⊕	⊕	⊕	⊕	⊕	⊕	⊕		
		⊕	⊕	⊕	⊕	⊕	⊕	⊕	⊕	⊕	⊕		

Figure 16.66 Multiple Skills Achievement Chart.

Production Management Boards

Application

Production management boards enable us to compare actual production results with the daily production schedule on an hourly basis, so that we can have early warning of scheduling problems (see Figures 16.67 and 16.68).

Main sections of form:

1. *Time.* Usually the time unit is one hour. Some factories may work better under a different time unit. Write the hours as start and finish times, such as "9:00 to 10:00."
2. *Standard output/total.* Enter the standard production output and total for the standard model.
3. *Date or model.* Usually, the horizontal boxes are for entering dates or the days of the week. Sometimes, the model name and model-specific pitch time are entered instead. [Example of date: 2/27 (Monday); Example of model: A113 (63 seconds)]
4. *Actual output/total.* Enter the actual output results and totals here.
5. *Reason.* If production lags behind schedule, enter the reason, such as "missing part" or whatever.

Period:	Production Management Board	Line:	Cycle time:
		Process:	
		Operation:	Standard no. operators:
		Previous process: / Next process:	By:

Time	Standard output / Total	Actual output / Total	Reason	Actual output / Total	Reason	Actual output / Total	Reason	Actual output / Total	Reason	Actual output / Total	Reason	Actual output / Total	Reason	Actual output / Total	Reason	Actual output / Total	Reason
8–9	60 / 60	58 / 58															
9–10	60 / 120	60 / 118															
2–3	60 / 365	60 / 365															
3–4	60 / 395																
4–5	60 / 495																
Results																	
Actual total	Differ. in totals																
Response to abnormalities																	

Figure 16.67 Example of Production Management Board.

Figure 16.68   Production Management Board.

Model and Operating Rate Trend Charts

Application

These charts (see Figures 16.69 and 16.70) help us understand what needs to be done during changeover operations. The model and operating rate trend chart shows how changing models affects changes in the operating rate. It can also be used to show the relationship between frequency of change-overs and operating rates.

Main sections of form:

1. *Model (or changeover frequency).* Set monthly averages for annual results and then select a month's figure as a sample. Enter the model changes (or changes in change-over frequency) here.
2. *Operating rate.* Enter how the operating rate changes when the model is changed or when the changeover frequency is changed.
3. *Chart area.* Make a bar graph based on the data for changes in model (or changeover frequency) and changes in the operating rate.

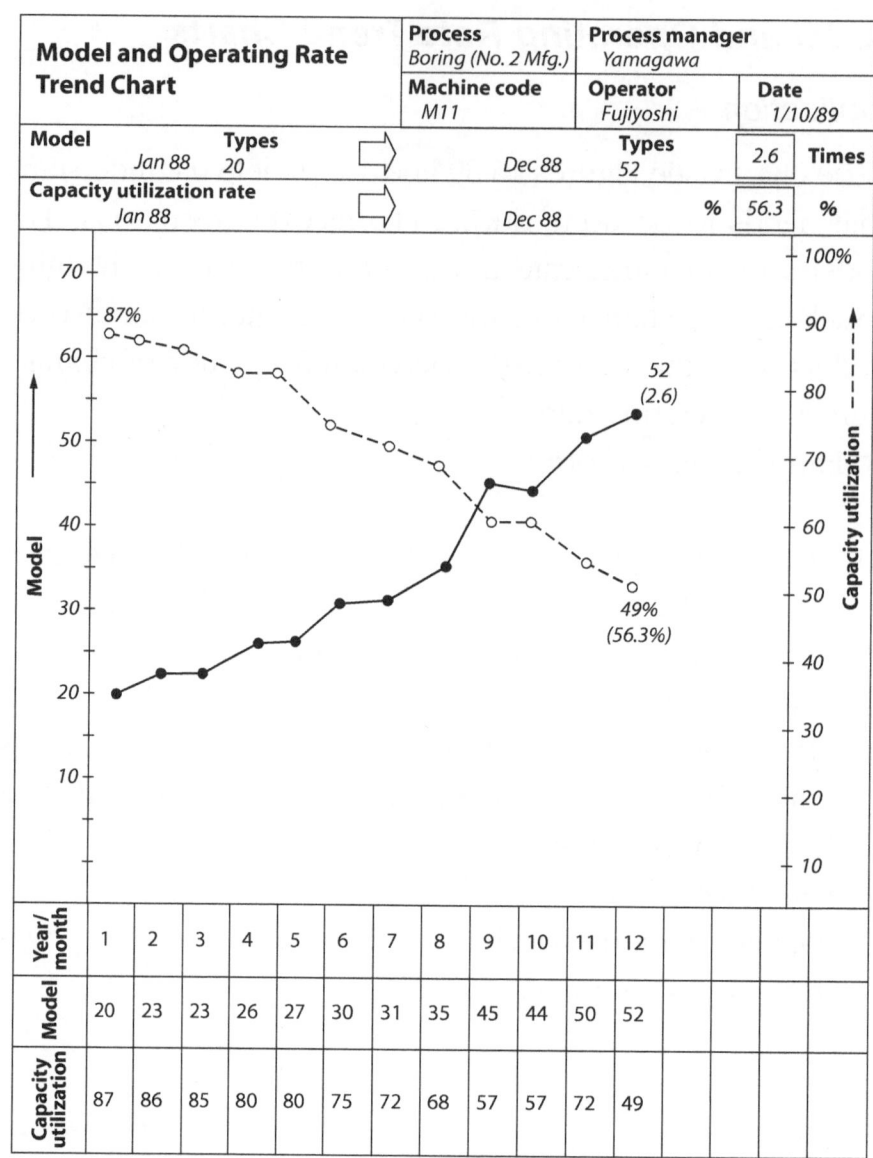

Figure 16.69   Example of Model and Operating Rate Trend Chart.

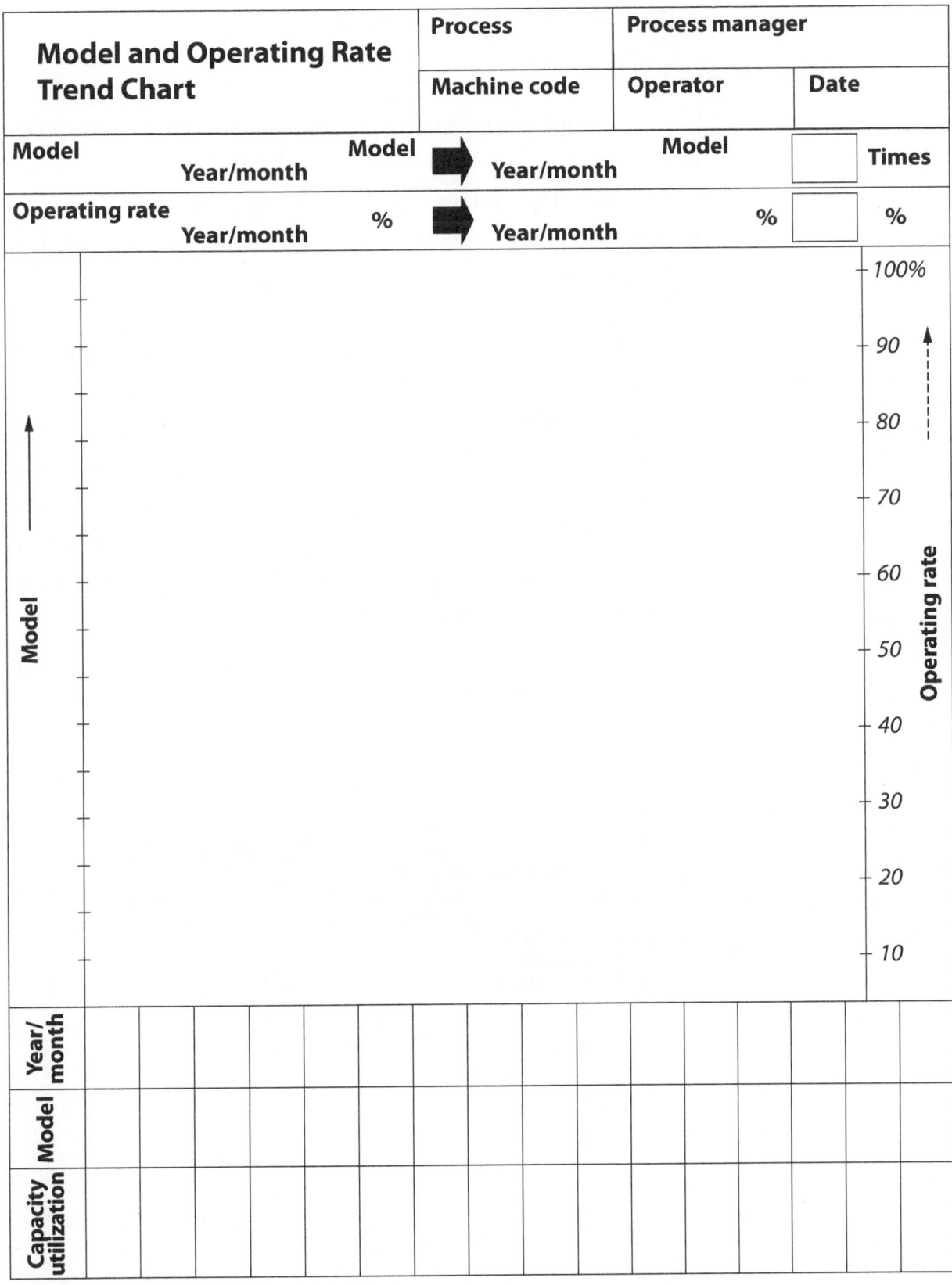

Figure 16.70   Model and Operating Rate Trend Chart.

Public Changeover Timetables

Application

Once an improved pattern is devised for a certain product model, the improvement team should perform a public changeover. This table (see Figures 16.71 and 16.72) helps make the improved operations and methods more explicit.

Main sections of form:

1. *Model.* Enter the name of the model being switched from and the one being switched to.
2. *Operator (timer).* Enter the name of the changeover operator and the name of the person who is timing the operator.
3. *Time.* Enter the time measurements in minutes and seconds.
4. *Operations.* For each operator, write on the form which operations were done at which point along the vertical time axis.

Public Changeover Timetable (1)	**Process (code)** No. 2 Mfg, No. 2 Print		**Description of changeover** Change colors on two rolls	
	Model AV1N400 (600")>MAGDP (400")		**Entered by** Sandler	**Date** 11/3/88
Operator Time	**Operator (timer)** Lennon (David)	McTighe (Naill)	Rosen (Cheryl)	

Figure 16.71 Example of Public Changeover Timetable.

Public Changeover Timetable ( )	Process (code)		Description of changeover	
	Model		Entered by	Date
Operator / Time	Operator (timer)			

Figure 16.72   Public Changeover Timetable.

Changeover Improvement Lists

Application

These lists are for writing down more detailed improvement ideas that develop from further study of results from change-over work procedure analysis charts and other improvement-related data (see Figures 16.73 and 16.74).

Main sections of form:

1. *Description of changeover.* Briefly describe the type of changeover.
2. *Model.* Enter the name of the model being switched from and the one being switched to.
3. *Improvement items.* Enter which items are in need of improvement.
4. *Improvement topic.* Use tables or diagrams to specifically describe the improvement being studied.
5. *Person in charge.* Enter the name of the improvement team leader.
6. *Deadline.* Enter the improvement deadline date.
7. *Confirmation.* Have the section chief or changeover team leader confirm the improvement results.

Changeover Improvement List ( )		Process (code) *No. 2 Mfg, No. 2 Print*	Description of changeover *Change colors on two rolls*		
		Model *AV1N400 (600")>MAGDP (400")*	Entered by *Sandler*	Date *11/3/88*	
No.	Items in need of improvement	Proposed improvement	Person in charge	Target date	Confirm-ation
1	Line up plate cylinder carts in U-shaped formation, and make this a rule	Make carts easier to move around	Jones	1/10	
2	Make wrapping paper tube placement a one-touch operation	Use velcro tape to enable one-touch operation	Jones	1/10	
3	Make plate cylinder replacement a one-worker job (currently two workers)	Develop specialized carts, use stoppers and other jigs or tools to change internal changeover into external changeover	Jones	1/10	
4	Disable backward motion of plate cylinder carts	Place carts below pull-out rolls	Jones	1/10	

Figure 16.73   Example of Changeover Improvement List.

Changeover Improvement List ( )	Process (code)		Description of changeover		
	Model →		Entered by	Date	
No.	Items in need of improvement	Proposed improvement	Person in charge	Target date	Confirm-ation

Figure 16.74 Changeover Improvement List.

Changeover Work Procedure Analysis Charts

Application

These charts (see Figures 16.75 and 16.76) are tools for change-over improvements. We use them to analyze and elucidate the contents of changeover operations and to gain a detailed understanding of each specific task in those operations so that we can more accurately eliminate the inherent waste.

Main sections of form:

1. *Description of changeover.* Briefly describe the type of changeover.
2. *Model.* Enter the name of the model being switched from and the one being switched to.
3. *Operator (timer).* Enter the name of the changeover operator and the name of the person who is timing the operator.
4. *Changeover operation.* Describe a specific task in the changeover operation. Example: tighten fastening bolts (4 bolts).
5. *Read time.* This is the time measurement that begins at the start of the changeover operations.
6. *Operation time.* The operation times are taken from the read time after the changeover operations are performed.
7. *Changeover categories.* Note whether each changeover task falls under the category of external changeover or internal changeover.
8. *Improvement plan.* Describe the improvement plans regarding specific changeover tasks.

Changeover Operations Analysis Chart	Minutes 39.7	**Process (code)** No. 3 Mfg, Press M13		**Description of changeover** Mold change and width adjustment		
		Model MC377-01>CB211.02		**Operator (timer)** Yamashima		**Date** 12/15/88

No.	Changeover operation	Read time	Operation time	Changeover categories			Improvement plan
				Internal	External	Waste	
1	Go to pick up tools	1'05"	1'05"		✓		Make specialized cart for picking up tools
2	Go to pick up lift set	3'15"	2'15"		✓		Make specialized cart for picking up lift sets
3	Remove cover bolts (4)	5'55"	2'40"	✓			Remodel for boltless design
4	Remove side bolts (6)	8'59"	3'04"	✓			

Figure 16.75   Example of Changeover Operations Analysis Chart (1).

Changeover Operations Analysis Chart	Minutes	Process (code)		Description of changeover			
		Model ⟶		Operator (timer)		Date	

No.	Changeover operation	Read time	Operation time	Changeover categories			Improvement plan
				Internal	External	Waste	

Figure 16.76 Changeover Operations Analysis Charts.

Changeover Results Table

Application

This table looks at actual changeover operations minute by minute to help us get a better idea of how those operations proceed (see Figures 16.77 and 16.78).

Main sections of form:

1. *Period.* Enter the period during which the changeover measurements were made.
2. *Model.* Enter the name of the model being switched from and the one being switched to.
3. *Operator.* Enter the name of the changeover operator.
4. *Changeover time.* Enter measured changeover times as a bar graph along the form's horizontal time axis.

Changeover Results Table (1)					Process name			Measurements by			
					No. 3 Mfg, Press			*Jones*			
					Machine code			Period			
					M13-123			*10/1/88 to 12/1/88*			

Item		Date	Operator	Change-over time	Time (minutes)				
Before	After				10	20	30	40	
CV311 –05	CA231 –06	10/1	Jones	15′ 30″					
CA231 –06	VA921 –04	10/1	″	20′ 10″					
VAR21 –04	MC379 –01	10/1	″	14′ 13″					
MC377 –01	CB211 –02	10/1	″	25′ 54″					
CB211 –02	NA366 –03	10/2	″	14′ 01″					
M366 –03	2N11 –01	10/3	″	18′ 24″					

Figure 16.77   Example of Changeover Results Table.

Changeover Results Table ( )		Process name	Measurements by				
		Machine code	Period				
Item		Date	Operator	Change-over time	Time (minutes)		
Before	After				10　　20　　30　　40		

Figure 16.78　Changeover Results Table.

5S Checklist for Changeover

Application

We use this list to check up on how well the 5S's are being maintained as a basic requirement for efficient changeover.

Main sections of form (see Figures 16.79 and 16.80):

1. *Checker.* Enter the name of the person checking the changeover operation.
2. *5S check point.* Enter the 5S check points for each machine involved in the changeover operation.
3. *Check column.* Enter the month and date of the check, the checker's initials, and the check symbols. This can be simply a check mark or, as the Japanese do, a circle for "good," a triangle for "OK," and an "X" for "not good."

		Process			Machine code					
Changeover 5S Checklist		No. 2 Mfg, boring line			MD001					
		Workshop leader			**Checked by**		**Date**			
		Rivera			Rivera		1/15/89			

No.	5S checklist item	Date and Operator	1/9	1/10	1/11	1/12	1/13	1/14	1/15	Improvement plan
1	Are different sets of jigs and tools used for changeover kept by each machine?		○	○						
2	Are the jigs and tools within easy reach during changeovers?		○	○						
3	Are the jigs and tools laid out according to the order of use during changeover?		○	○						
4	Are the jigs and tools laid out in an orderly manner?		○	○						
5	Are there some carts reserved expressly for use in changeover and do they have a prescribed storage site?		○	○						
6	Are the items in the carts arranged in an orderly manner?		○	△						
7	Does each machine carry instructions from the changeover operations manual?		△	△						
8	Are the operators performing the changeover as instructed in the manual?		✕	✕						
9	Have quality standards been set for each model?		○	○						
10	Are the standards posted on each machine?		○	○						
11	Are defect-free samples of each model on display for reference?		○	○						

Figure 16.79 Example of Changeover 5S Checklist.

Changeover 5S Checklist		Process		Machine code							
		Workshop leader		Checked by			Date				
No.	5S checklist item	Date and Operator									Improvement plan

Figure 16.80 Changeover 5S Checklist.

Poka-Yoke/Zero Defects Checklist

Application

This checklist (see Figures 16.81 and 16.82) helps us find defect causes that arise from human error and helps elucidate the types, sequence, and progress of responses made to eliminate those causes.

Main sections of form and procedure for filling out form:

Poka-Yoke/Zero Defects Checklist		Division *Manufacturing*		Process *Door process*		Date 4/7/89		No.		
		Department *Prep.*		Machine model		Entered by *Hirada*				

#	Operation	Operation (machine)	Standard	Inspection	Defect description	Defect cause	3-point evaluation				3-point response				Description of response (evaluation)	Deadline	Person in charge
							Occurrence frequency	Impact on processes	Impact on company	Total points	Urgency	Difficulty	Countermeasure	Total points			
1	Remove workpiece	Manual operation		None	Damaged	Workpieces rubbed	2	2	2	8	1	2	2	4	Store workpieces so they don't rub	5/20	Ozaki
		S101 lifter			Dented	Workpieces collided	1	2	2	4	1	2	2	4			
2	S101 small groove process	Manual operation	2×600	None	Defective groove width	Error in cutting	1	2	2	4	1	2	2	4			
					Bent groove	Wrong jig	2	2	2	8	2	2	2	8	Use limit switch to check length	5/20	Ozaki
3	Store workpiece			None	Damaged	Workpieces rubbed	2	2	2	8	1	2	2	4	Store workpieces so they don't rub	5/20	Ozaki
					Dented	Workpieces collided	1	2	2	4	1	2	2	4			

Figure 16.81   Example of *Poka-Yoke*/Zero Defects Checklist.

#	Operation	Operation (machine)	Standard	Inspection	Defect description	Defect cause	3-point evaluation				3-point response				Description of response (evaluation)	Deadline	Person in charge
							Occurrence frequency	Impact on processes	Impact on company	Total points	Urgency	Difficulty	Countermeasure	Total points			

Poka-Yoke/Zero Defects Checklist — Division, Department, Process, Machine model, Date, No., Entered by

Figure 16.82 *Poka-Yoke*/Zero Defects Checklist.

Parts-Production Capacity Work Table

Application

This is a tool for promoting standard operations. It clarifies the basic time and processing capacity for each part that is processed (see Figures 16.83 and 16.84).

Main sections of form:

1. *Serial number.* Enter the serial number of processing machine being used.
2. *Manual operation time.* Enter the amount of manual operation time in minutes and seconds.
3. *Auto feed time.* Enter the amount of machine processing time (after starting the machine) in minutes and seconds.
4. *Completion time.* This is the sum of the manual operation time and the auto feed time. If doing parallel operations, show the parallel times rather than the total.

    Example: If the manual operation time is five seconds and the auto feed time is 25 seconds, but the two are done in parallel, show them as overlapping graph times.
5. *Per-unit changeover time.* To get this figure, divide the total changeover time by the number of units exchanged.
6. *Graph time indications.* Show graph times as serial or overlapping.

Approval stamps		Parts-Production Capacity Work Table		Part No.		Type RY		Entered by Sato	
				Part name 6" pinion		Quantity 1		Creation date 1/17/89	

Process	Process name	Serial No.	Manual operation time (A)		Basic times — Auto feed time (B)		Complet-ion time C=A+B		Blades and bits — Retooling amount (D)	Retooling time (E)	Per unit retooling time F=E+D	Total time per unit G=C+F	Production capacity I/G	Graph time — Manual work - - - Auto feed ———
			Min.	Sec.	Min.	Sec.	Min.	Sec.						
1	Pick up raw materials	——		1		—		1	——	——	——	1	——	
2	Gear teeth cutting	A01		4		35		39	400	2'10"	0.3"	39.3	717	4" - 35"
3	Gear teeth surface fin.	A02		6		15		21	1,000	2'00"	0.1"	21.1	1,336	6" - 15"
4	Foward gear surface fin.	A03		7		38		45	400	3'00"	0.5"	45.5	619	7" - 38"
5	Reverse gear surface fin.	A04		5		28		33	400	2'30"	0.4"	33.4	844	5" - 28"
6	Pin width measurement	B01		8		5		13	——	——		13	259	8" 5"
7	Store finished workpiece	——		1		—		1	——	——	——	1	——	

Figure 16.83 Example of Parts-Production Capacity Work Table.

Approval stamps		**Parts-Production Capacity Work Table**		Part No.		Type		Entered by		
				Part name		Quantity		Creation date		
Process	Process name	Serial No.	Manual operation time (A)	Basic times		Blades and bits		Per unit retooling time F = E/D	Total time per unit G = C+F	Production capacity I/G	Graph time
				Auto feed time (B)	Complet-ion time C = A+B	Retooling amount (D)	Retooling time (E)				Manual work —— Auto feed - - - - -
			Min.	Sec.	Min.	Sec.	Min.	Sec.				

Figure 16.84   Parts-Production Capacity Work Table.

Standard Operations Combination Chart

Application

This chart is an analytical tool that helps us find out just how people and machines combine their labor during operations so that we can find a more efficient combination.

Main sections of form (see Figures 16.85 and 16.86):

1. *Required output.* Enter the required output per day.
2. *Cycle time.* Enter the per-unit cycle time, calculated as the total operating time divided by the required output.
3. *Manual labor.* Enter the amount of manual labor time.
4. *Auto feed.* Enter the amount of automated labor (auto feed) time.
5. *Walking.* Enter the amount of walking time between processes.
6. *Form completion method.* Indicate manual labor as solid lines, auto feed as broken lines, and walking as wavy lines, as shown in Figure 16.85.

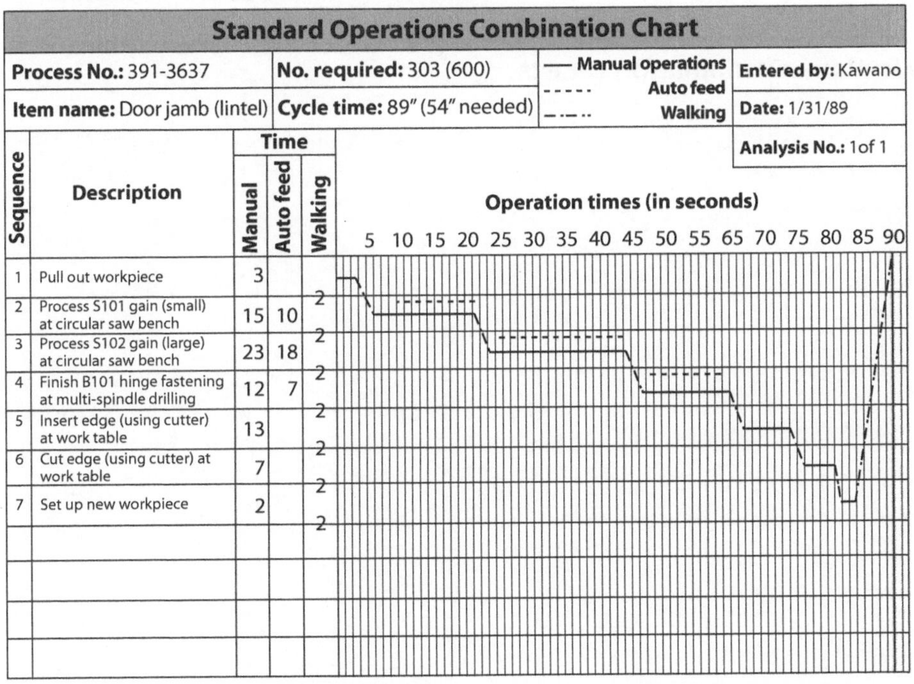

Figure 16.85   Example of Standard Operations Combination Chart.

Standard Operations Combination Chart

| Process No.: | | | No. required: | | | —— Manual operations | Entered by: |
| Item name: | | | Cycle time: | | | - - - - Auto feed
—·—·· Walking | Date: |

Sequence	Description	Time			Operation times (in seconds)	Analysis No.:
		Manual	Auto feed	Walking	5 10 15 20 25 30 35 40 45 50 55 60 65 70 75 80 85	

Figure 16.86   Standard Operations Combination Charts.

Summary Table of Standard Operations

Application

This table should include descriptions of all the essential components of production operations, such as operating equipment, exchanging jigs, changeover, processing procedures, and so on. When completed, it can be a useful tool for training new workers (see Figures 16.87 and 16.88).

Main sections of form:

1. *Processing sequence.* Enter the processes in the order of their execution.
2. *Machine number.* Enter the number (serial number, etc.) of the machine being used.
3. *Description of operation.* Describe what happens in the operation and include the key procedural points.
4. *Critical factors.* Describe all critical factors in areas such as correct methods, incorrect methods, safety, quality, and so on.
5. *Diagram.* Draw a simple diagram of the operation.

Summary Table of Standard Operations		Process name	Department	Date	Confirmation			
		Processing sequence						
		Machine number						
No.	Description of operation	Critical factors (correct/incorrect, safety, facilitation, etc.)	Diagram of operation					

Figure 16.87 **Example of Summary Table of Standard Operations.**

Summary Table of Standard Operations	Process name		Department	Date	Confirmation			
	Processing sequence							
	Machine number							
No.	**Description of operation**	**Critical factors** (correct/incorrect, safety, facilitation, etc.)	**Diagram of operation**					

Figure 16.88   Summary Table of Standard Operations.

Work Methods Table

Application

This table instructs workers in the standard operations for each process (see Figures 16.89 and 16.90). It can serve as a useful guide for workers being trained in multiple skills.

Main sections of form:

1. *Description of operation.* Give a specific description of the methods used in the operation.
2. *Quality.* Describe the quality checking procedures, measurement methods, and so on.
3. *Critical factors.* Describe all critical factors in areas such as correct methods, incorrect methods, safety, quality, and so on.
4. *Net time.* Enter the net time for the operation (exclusive of quality checks, changeover, and other peripheral tasks).
5. *Diagram.* Draw a simple diagram of the operation, as in the summary table of standard operations.

Work Methods Table		Part no.	Required output	Dept.	Name	Confirmation		
		Part name	Breakdown no.		Date			

No.	Description of operation	Quality		Critical factors (correct/incorrect, safety, facilitation, etc.)	Net time (min. and sec.)	Cycle time	Stand. in-process inv.	Stand. in-process inv.	Safety point	Quality check point
		Check	Measure.					●	✚	◇

Figure 16.89   Example of Work Methods Table.

Work Methods Table		Part no.		Required output	Dept.	Name	Confirmation
		Part name		Breakdown no.		Date	

No.	Description of operation	Quality		Critical factors (correct/incorrect, safety, facilitation, etc.)	Net time (min. and sec.)	Cycle time	Stand. in-process inv.	Stand. in-process inv. ●	Safety point ✚	Quality check point ◇	
		Check	Measure.								

Figure 16.90   Work Methods Table.

Standard Operations Form

Application

Use this form to provide a visual description of the equipment layout, cycle time, work sequence, standard in-process inventory, and other critical factors in the correct execution of standard operations (see Figures 16.91 and 16.92).

Main sections of form:

1. *Cycle time.* Enter the cycle time indicated in the standard operations combination chart. The (per-unit) cycle time is obtained by dividing the total operating time in a day by the required output for that day.
2. *Net time.* This is the minimum time required to execute the operation, exclusive of all quality checks, changeover, and other peripheral tasks.
3. *Number of standard in-process inventory points.* Use shaded circles to indicate each instance of standard in-process inventory. Enter one shaded circle for each auto-feed machine and another one for each reverse-order operation.
4. *Safety points.* Use solid crosses to indicate safety points, such as for tasks involving the blade or bit exchanges, and so on.

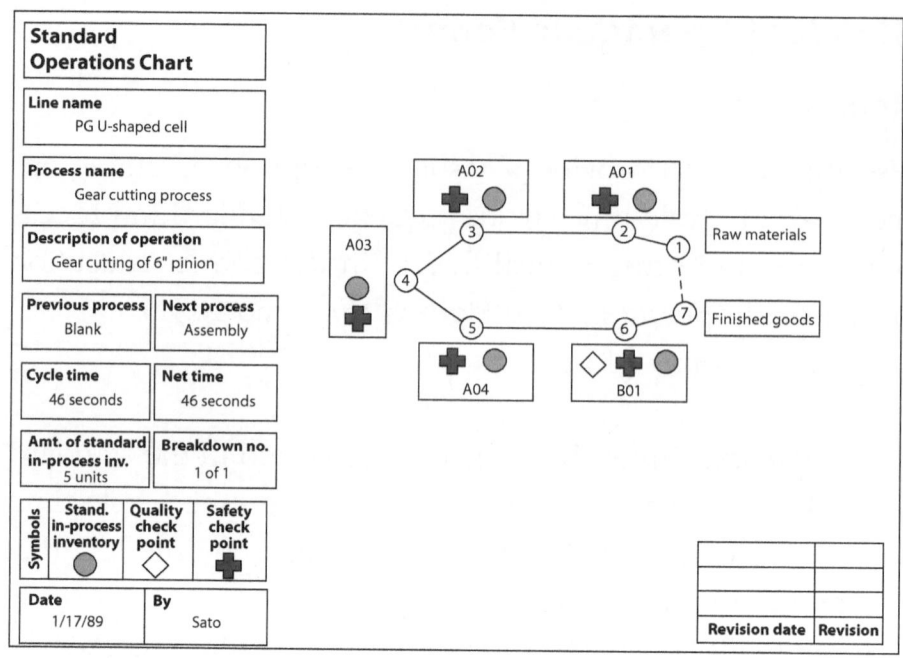

Standard Operations Chart	

Line name
PG U-shaped cell

Process name
Gear cutting process

Description of operation
Gear cutting of 6" pinion

Previous process	**Next process**
Blank	Assembly

Cycle time	**Net time**
46 seconds	46 seconds

Amt. of standard in-process inv.	**Breakdown no.**
5 units	1 of 1

Symbols	**Stand. in-process inventory**	**Quality check point**	**Safety check point**

Date	**By**
1/17/89	Sato

Revision date	**Revision**

Figure 16.91   Example of Standard Operations Form.

Standard Operations Chart	
Line name	
Process name	
Description of operation	
Previous process	**Next process**
Cycle time	**Net time**
Amt. of stand. in-process inv.	**Breakdown no.**

Symbols	Stand. process inventory ⬤	Quality check point ◇	Safety check point ✚

Date	By

Revision date	**Revision**

Figure 16.92   Standard Operations Form.

JIT Introduction-Related Forms

JIT's Ten Commandments

Application

The JIT factory revolution is a battle that companies wage against themselves (at least, against their "old" selves). To grow and change with the times, companies must break down the edifices of their old habits and wholeheartedly undertake improvements. This revolutionary spirit must permeate every part of the company, from the top managers' offices to the JIT promotional headquarters, clerical centers, and every workshop in the factory (see Figure 16.93).

It might be a good idea to have the following list of "JIT's Ten Commandments" on display at every improvement meeting.

Improvement Memos

Application

This memo is a simple form to be used for passing along pointers, improvement ideas, and whatever else might further the cause of improvement. The bottom part includes category boxes to aid in filing and managing the memos.

Main sections of form (see Figure 16.94):

1. *Person in charge.* Enter the name of the person in charge of making the improvement.
2. *Deadline.* Enter the improvement deadline date.
3. *Pointer.* Enter whatever pointers have been given by an improvement leader or JIT consultant.
4. *Description of improvement.* Briefly describe the improvement pertaining to the pointer.
5. *JIT function categories.* Check off the box for the JIT function that the memo relates to, such as the 5S's, flow production, and so on.

JITs "Ten Commandments"

1. Throw out traditional concepts of manufacturing methods.

2. Think of how the new method will work—not how it won't.

3. Don't accept excuses. Totally deny the status quo.

4. Don't seek perfection. A 50-percent implementation rate is fine as long as it's done on the spot.

5. Correct mistakes the moment they're found.

6. Don't spend money on improvements.

7. Problems give you a chance to use your brain.

8. Ask "Why?" five times.

9. Ten person's ideas are better than one person's.

10. Improvement knows no limits.

Figure 16.93   JIT's Ten Commandments.

Improvement memo								
Person in charge				**Deadline**				
K. Jones, 4th floor				3/6/90				
Pointer								
Talk to 3rd floor office supply person.								
Description of improvement								
Label shelves in office supply cabinet.								
☐	☐	☐	☑	☐	☐	☐	☐	☐
5S	Flow production	Multi-process operations	Labor cost reduction	Visual control	Leveled production	Jidoka	Other	

Figure 16.94   Example of Improvement Memo.

List of JIT Improvement Items

Application

Use this form to list improvement ideas presented on improvement memos to help gauge the progress of improvement activities (see Figures 16.95 to 16.97).

Main sections of form:

1. *Improvement item.* Enter the name of the item undergoing improvement.
2. *Person in charge.* Enter the name of the person in charge of the above improvement item.
3. *Start date.* Enter the date when this improvement campaign began.
4. *Deadline.* Enter the improvement deadline date.
5. *Confirmation.* Have the section chief or JIT office representative confirm the progress or completion of the improvement campaign by the deadline date.

Figure 16.95　Improvement Memo.

List of JIT Improvement Items	Section or group:				
	By:		Date:		
No.	**Improvement item**	**Person in charge**	**Start date**	**Deadline**	**Confirmation**

Figure 16.96 Example of List of JIT Improvement Items.

List of JIT Improvement Items	Section or group:				
	By:			Date:	
No.	Improvement item	Person in charge	Start date	Deadline	Confirmation

Figure 16.97   List of JIT Improvement Items.

Improvement Campaign Planning Sheet

Application

During large or long-term improvement campaigns, we sometimes need to stop in the middle to gauge our progress to date and estimate whether we will be able to complete the improvement by the deadline. This planning sheet provides a handy form for gathering the information required for such mid-point evaluations (see Figures 16.98 and 16.99).

Main sections of form:

1. *Theme.* Enter the improvement theme and improvement item(s).
2. *Before and after diagrams.* Describe in text and/or diagrams the situation before improvement and the intended situation when the improvement is completed.
3. *Implementation items.* Describe in detail the specific improvements being made as part of the improvement theme.
4. *Time scale.* Enter the schedule area's time scale as a four-month period. Indicate ten-day intervals.
5. *Schedule area.* Fill out the schedule area with broken lines in the upper part of each row to indicate the estimated scheduling, and solid lines in the lower part to indicate the actual schedule.
6. *Problems and future considerations.* Jot down brief descriptions of any problems that have arisen, as well as topics for future study.
7. *Impact.* Enter the estimated impact if the actual impact is not yet known.

Improvement Campaign Planning Sheet		Department		
		Person in charge S. Ott		Date 4/1/90

Theme

Before Improvement	After Improvement
1. ⎯⎯⎯⎯	
2. ⎯⎯⎯⎯	
3. ⎯⎯⎯⎯	

No.	Implementation items	Person in charge	Time scale 10 20 10 20 10 20 10 20	Comments
1.	⎯⎯⎯	KJ		
2.	⎯⎯⎯	DL		
3.	⎯⎯⎯	MA		

Problems and future considerations	Impact

Figure 16.98   **Example of Improvement Campaign Planning Sheet.**

Improvement Campaign Planning Sheet	Department		
	Person in charge		Date

Theme

Before Improvement	After Improvement

No.	Implementation items	Person in charge	Time scale 10 20 10 20 10 20 10 20	Comments

Problems and future considerations	Impact

Figure 16.99   Improvement Campaign Planning Sheet.

Improvement Results Charts

Application

After an improvement is completed, it is a good idea to use these forms to create a display that shows before and after photographs of the improvement site and basic data, such as the costs incurred by the improvement, the impact of the improvement, and the like (see Figures 16.100 and 16.101).

Main sections of form:

1. *Before improvement.* Attach a photograph of the improvement site before improvement.
2. *After improvement.* Also attach a photo of the improvement site after improvement.
3. *Problem points.* Briefly describe the problem points addressed by the improvement.
4. *Improvement points.* Briefly describe the main improvement points.
5. *Costs.* Summarize the costs incurred by the improvement.
6. *Impact.* Describe the actual impact of the improvement.

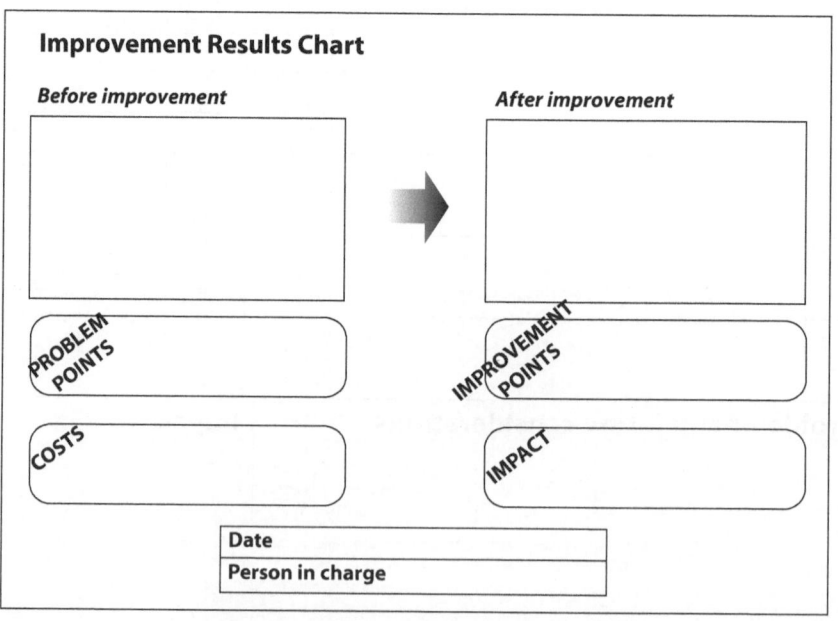

Figure 16.100   Example of Improvement Results Chart.

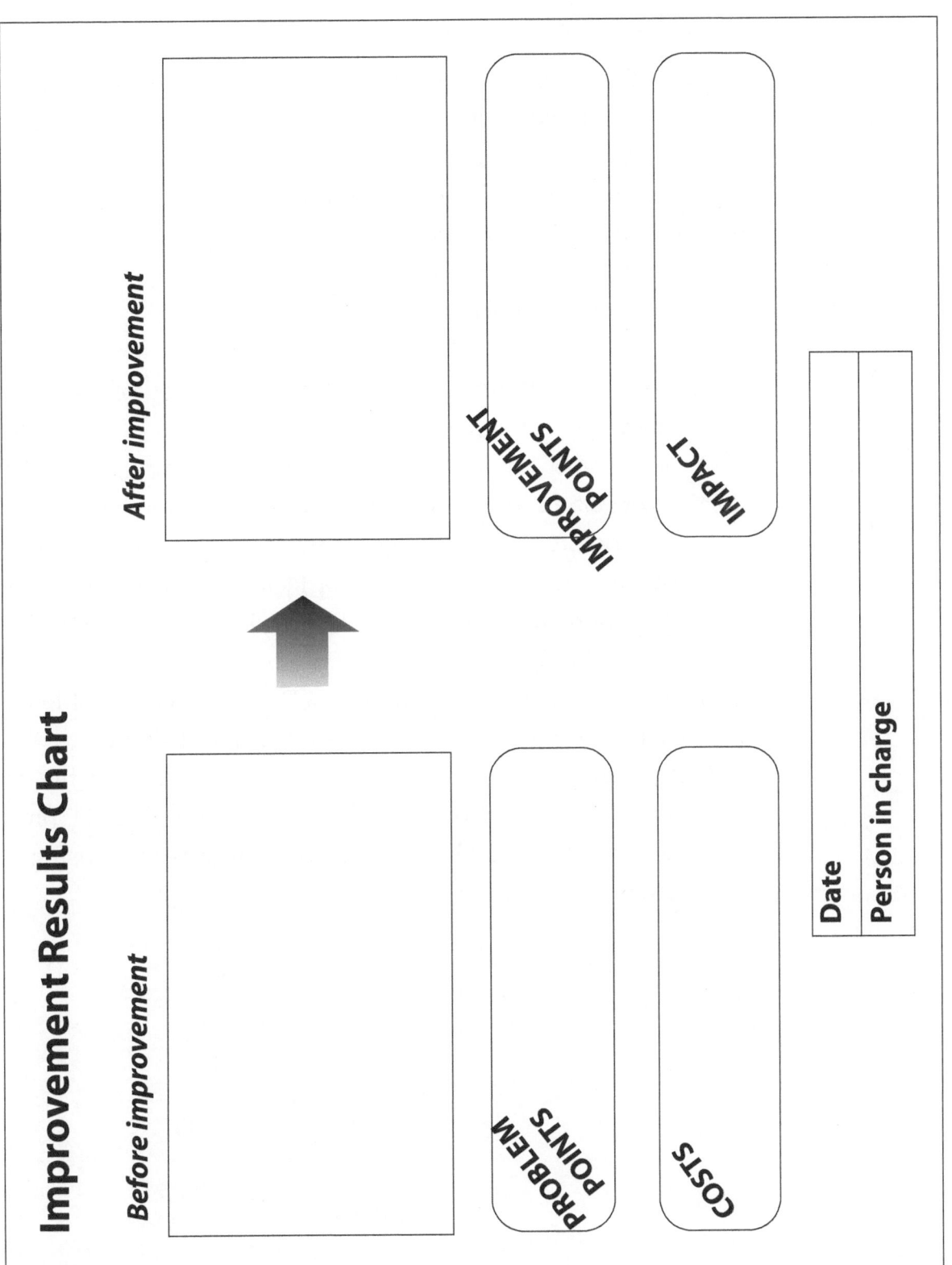

Figure 16.101 Improvement Results Chart.

Weekly Report on JIT Improvements

Application

When subcontractors, subsidiaries, or factories carry out their own JIT improvement activities, they should be encouraged to fill out one of these report forms every week to keep the JIT promotional headquarters informed of their progress and of the types of improvement measures being undertaken (see Figures 16.102 and 16.103).

Main sections of form:

1. *Name of company or factory.* Enter the name of the sub-contractor, subsidiary, or factory.
2. *Month/date.* Enter the current month and date.
3. *Improvement items.* Describe the improvement items that were pointed out to the company or factory, or that were established by the company or factory themselves.
4. *Description of improvement.* Briefly describe the improvement(s) being made.
5. *Problems and countermeasures.* Describe what problems have arisen and what is being done about them.
6. *Impressions.* The relevant supervisor should enter his or her impression of the improvement activity The JIT promotion office may later add its own impressions, using a different color ink.

Weekly Report on JIT Improvement		Company or factory:	
		By:	
Date / (Day of week)	**Improvement items**	**Description of improvement**	**Problems/ countermeasures**
3/6 (MON)	1.		
3/7 (TUE)	2.		
3/8 (WED)	3.		
3/9 (THU)			
3/10 (FRI)			
Impressions			

Figure 16.102   Example of Weekly Report on JIT Improvements.

Weekly Report on JIT Improvement			Company or factory:
			By:
Date / (Day of week)	**Improvement items**	**Description of improvement**	**Problems/ countermeasures**
/ (MON)			
/ (TUE)			
/ (WED)			
/ (THU)			
/ (FRI)			
Impressions			

Figure 16.103   Weekly Report on JIT Improvements.

JIT Leader's Report

Application

Intended mainly for use by head-office industrial engineers and outside consultants, JIT leaders may find this form useful when providing guidance to factories, subsidiaries, or subcontractors (see Figures 16.104 and 16.105).

Main sections of form:

1. *JIT leader's name.* Enter the name of the industrial engineer or JIT consultant.
2. *To:* Enter the company or factory name and the department or division.
3. *Person in charge.* Enter the name of the person in charge at the company or factory.
4. *Advice.* Explain the point of the report, including methodology
5. *Conditions.* Describe any obstacles or other adverse conditions at the company or factory relevant to the advice given.
6. *Outlook.* Describe the outlook for the company or factory.
7. *Problems and solutions.* Point out the existing problems and suggest solutions.

JIT Leader's Report	**Date:** 4/4/90
	JIT leader's name: D. Lennon
To:	**Consultation date:** 4/2/90
Person in charge: S. Ott	
Advice: 1.	

Figure 16.104   Example of JIT Leaders' Report.

JIT Leader's Report	Date:
	JIT leader's name:
To:	Consultation date:

Person in charge:
Advice:
Conditions
Outlook
Problems and solutions

Figure 16.105   JIT Leaders' Report.

Index

B

K

S

Y

Z

About the Author

Hiroyuki Hirano believes Just-In-Time (JIT) is a theory and technique to thoroughly eliminate waste. He also calls the manufacturing process the equivalent of making music. In Japan, South Korea, and Europe, Mr. Hirano has led the on-site rationalization improvement movement using JIT production methods. The companies Mr. Hirano has worked with include:

Polar Synthetic Chemical Kogyo Corporation
Matsushita Denko Corporation
Sunwave Kogyo Corporation
Olympic Corporation
Ube Kyosan Corporation
Fujitsu Corporation
Yasuda Kogyo Corporation
Sharp Corporation and associated industries
Nihon Denki Corporation and associated industries
Kimura Denki Manufacturing Corporation and associated industries
Fukuda ME Kogyo Corporation
Akazashina Manufacturing Corporation
Runeau Public Corporation (France)
Kumho (South Korea)
Samsung Electronics (South Korea)
Samsung Watch (South Korea)
Sani Electric (South Korea)

Mr. Hirano was born in Tokyo, Japan, in 1946. After graduating from Senshu University's School of Economics, Mr. Hirano worked with Japan's largest computer manufacturer in laying the conceptual groundwork for the country's first full-fledged production management system. Using his own interpretation of the JIT philosophy, which emphasizes "ideas and techniques for the complete elimination of waste," Mr. Hirano went on to help bring the JIT Production Revolution to dozens of companies, including Japanese companies as well as major firms abroad, such as a French automobile manufacturer and a Korean consumer electronics company.

The author's many publications in Japanese include: *Seeing Is Understanding: Just-In-Time Production* (*Me de mite wakaru jasuto in taimu seisanh hoshiki*), *Encyclopedia of Factory Rationalization* (*Kojo o gorika suru jiten*), *5S Comics* (*Manga 5S*), *Graffiti Guide to the JIT Factory Revolution* (*Gurafiti JIT kojo kakumei*), and a six-part video tape series entitled *JIT Production Revolution, Stages I and II*. All of these titles are available in Japanese from the publisher, Nikkan Kogyo Shimbun, Ltd. (Tokyo).

In 1989, Productivity Press made Mr. Hirano's *JIT Factory Revolution: A Pictorial Guide to Factory Design of the Future* available in English.